To: Helen & Cla[illegible]

My first Autographed Book

Ron [illegible]

MOONSHINE MARKETS

International Center for Alcohol Policies
Series on Alcohol in Society

Grant & Litvak—*Drinking Patterns and Their Consequences*
Grant—*Alcohol and Emerging Markets: Patterns, Problems, and Responses*
Peele & Grant—*Alcohol and Pleasure: A Health Perspective*
Heath—*Drinking Occasions*
Houghton & Roche—*Learning About Drinking*
Haworth & Simpson—*Moonshine Markets: Issues in Unrecorded Alcohol Beverage Production and Consumption*

MOONSHINE MARKETS

Issues in Unrecorded Alcohol Beverage Production and Consumption

Edited by
Alan Haworth and Ronald Simpson

Brunner-Routledge
New York and Hove

Published in 2004 by
Brunner-Routledge
29 West 35th Street
New York, NY 10001
www.brunner-routledge.com

Published in Great Britain by
Brunner-Routledge
27 Church Road
Hove, East Sussex
BN3 2FA
www.brunner-routledge.co.uk

Brunner-Routledge is an imprint of the Taylor & Francis Group.
Printed in the United States of America on acid-free paper.

10 9 8 7 6 5 4 3 2 1

Library of Congress Cataloging-in-Publication Data

Moonshine markets : issues in unrecorded alcohol beverage production and consumption / edited by Alan Haworth and Ronald Simpson.
p. cm. — (International center for alcohol policies series on alcohol in society)
Includes bibliographical references and index.
ISBN 0-415-93547-4 (Hardback)
1. Distilling, Illicit. 2. Drinking of alcoholic beverages--Health aspects. 3. Drinking of alcoholic beverages--Economic aspects. I. Haworth, Alan. II. Simpson, Ronald. III. Series on alcohol in society.
HJ5021.M66 2003
339.4'86631--dc21

2003012748

Contents

Editors

Alan Haworth, MD

Alan Haworth has been professor of psychiatry in the University of Zambia since 1977 and prior to this worked at developing a countrywide mental health service in Zambia from 1964. He began his studies of drinking behavior in 1969 when examining the provision of primary health care and in 1976 joined a World Health Organization (WHO) collaborative study of alcohol-related problems in three countries (Mexico, Scotland, and Zambia). He has worked as a consultant on alcohol-related problems for WHO and for the British Commonwealth Health Secretariat in seven African countries and has contributed chapters on drinking in Africa to a number of books. He is currently collecting preliminary data for a study of drinking in relation to HIV infection in Zambia.

Ronald C. Simpson, PhD

Dr. Simpson is a consultant to ICAP. During the past 25 years he worked in senior management positions in the food and beverage industry. Prior to retirement Dr. Simpson was Vice President of Corporate Scientific Affairs at Joseph E. Seagram and Sons for 10 years. He was responsible for developing and implementing strategies to gain a better understanding of the role of alcohol consumption in health and social issues. Dr. Simpson recieved his doctorate in Nutrition at the University of California at Davis.

Contributors

Dr. Linda A. Bennett
Department of Anthropology
University of Memphis
Memphis, Tennessee, USA

Anita Chopra
All India Institute of Medical Sciences
New Delhi, India

Dr. Gaurish Gaunekar
Institute of Psychiatry and Human Behavior
Goa, India

Marcus Grant
International Center for Alcohol Policies
Washington, D.C., USA

Dr. Alan Haworth
University of Zambia
Lusaka, Zambia

Dr. Nora Margaret Hogan
Muhimbili University College of Health Sciences
Dar es Salaam, Tanzania

Eleni Houghton
International Center for Alcohol Policies
Washington, D.C., USA

Dr. K. S. Jacob
Christian Medical College
Vellore, India

Dr. Navneet Johari
BJ Medical College
Ahmedabad, India

Dr. Gad Paul Kilonzo
Muhimbili University College of Health Sciences
Dar es Salaam, Tanzania

Dr. Kajiru Kilonzo
Muhimbili University College of Health Sciences
Dar es Salaam, Tanzania

Prof. Vladimir Nuzhnyi
Research Institute on Addictions
Ministry of Health
Moscow, Russia

Bertha Mamuya
Muhimbili University College of Health Sciences
Dar es Salaam, Tanzania

Dr. Jesse Mbwambo
Muhimbili University College of Health Sciences
Dar es Salaam, Tanzania

Dr. Davinder Mohan
All India Institute of Medical Sciences
New Delhi, India

Dr. Vikram Patel
Sangath Society
Goa, India
and
London School of Hygiene and Tropical Medicine
London, United Kingdom

Dr. Surajeen Prasad
Christian Medical College
Vellore, India

Dr. Anil Rane
Institute of Psychiatry and Human Behaviour
Goa, India

Dr. Haydée Rosovsky
Mexican Institute of Psychiatry
Mexico City, Mexico

Dr. Ronald Simpson
Consultant
U. Saddle River, New Jersey, USA

Dr. Eric Single
University of Toronto
Toronto, Ontario, Canada

Dr. Magda Vaissman
Institute of Psychiatry of the Federal University of Rio de Janeiro
and
Coordinator of Post Graduate courses on Chemical Dependence
Estacio de Sá University
Rio de Janeiro, Brazil

Prof. Ganpat Vankar
BJ Medical College
Ahmedabad, India

Prof. Grigory Zaigraev
All Russian Research Institute
Ministry of Internal Affairs
Moscow, Russia

Foreword

Moonshine Markets is the sixth book in the ICAP Series on Alcohol in Society. It explores consumption patterns of a type of alcohol that is generally, but not exclusively, illegal in most of the countries where it is consumed.

The book is an outgrowth of a meeting the International Center for Alcohol Policies (ICAP)* hosted in 1999, which brought together public health experts from developing countries to identify research topics of particular interest to these countries. Among the ideas that emerged was to study patterns of consumption of "local alcohol." Dr. Alan Haworth, a coeditor of this volume and a participant at the meeting, devised a methodology based on the completion of diaries in which participants described where, what, with whom, and how much alcohol they consumed over a specific 30-day period. The methodology was piloted in Tanzania and Zambia and then expanded to include Brazil, India, Mexico, and Russia.

Russia, although not a developing country, was included because of the high consumption of *samogon* in the rural areas and because the Russian Ministry of Health's Research Institute on Addictions (RIA) had a strong interest in this area. Indeed, ICAP is most grateful to the RIA for hosting the investigators' meeting in 2002, in which the research was reviewed and the current book took shape.

Our hope is that this volume will be of interest to government, the alcohol beverage industry, the alcohol research community, and indeed anyone concerned about alcohol, its history, and its patterns of consumption. Each of the country studies gives the reader a sense of the variety of products being consumed and the context in which the drinking takes place. This context includes both the actual drinking environment and the broader socioeconomic issues that may affect the way that beverages are consumed. The overarching

*ICAP is dedicated to helping reduce the abuse of alcohol worldwide and to promoting understanding of the role of alcohol in society through dialogue and partnerships involving the beverage alcohol industry, the public health community, and others interested in alcohol policy. ICAP is a not-for-profit organization supported by 10 major international beverage alcohol companies. This book was commissioned by ICAP. For more information on ICAP and its Cooperation Guidelines, visit www.icap.org.

chapters assess the evidence presented from an anthropological, economic, and health perspective.

Moonshine Markets explores an area of alcohol studies that is largely uncharted, but immensely important. The topic has been touched on in other books within the ICAP Series on Alcohol in Society, as well as through key informant surveys and some single-country studies,* but no cross-national study has been attempted to our knowledge. Although it is likely that "local alcohol" (unrecorded, illicit, noncommercial, or moonshine—however one chooses to refer to it) represents at least half of total alcohol consumption worldwide, we know very little about it. Despite the relatively small sample sizes in the country studies presented here, this book represents an unprecedented beginning in this field of cross-cultural inquiry and raises important issues that surprised even the investigators.

For example, it is commonly believed that the alcohol described in *Moonshine Markets* carries enormous health risks. Headlines from newspapers often cite deaths due to alcohol poisonings—in India, Kenya, Mexico, and South Africa, to cite just a few examples in recent years. In contrast to this perception, when a number of samples were tested from the drinks consumed by the study participants in the countries described in this book, Dr. Nuzhnyi (see chapter 11) found just the opposite. The samples contained no toxic properties and some were of quite high quality.

Dr. Nuzhnyi's analysis suggests that the quality of moonshine may not always be as injurious to health as has previously been assumed. Indeed, the notorious cases of alcohol poisoning may result from "bad batches" of moonshine, possibly produced by unscrupulous individuals, with no interest beyond immediate profit from the batch in question. Thus, the lack of quality control may be as important a determinant of harm as the pattern in which the products are consumed. Given the sheer quantities of moonshine consumed worldwide, the patterns of consumption of these products may be a pressing priority. This book begins a process that is worthy of further study.

In most of ICAP's books in this series, there is a chapter authored by a representative of the alcohol beverage industry. ICAP strongly believes that the industry, government, and research should all be represented in debates about alcohol issues. In *Moonshine Markets*, however, we have consciously elected not to include an industry perspective. In one sense, asking for such input on this topic is a little like asking people who pay taxes how they feel about people who do not. In a more general sense, there was little to add that would not have been discussed already by other contributors.

The editors and contributors of *Moonshine Markets* have also addressed the health policy implications of "local alcohol." Indeed, the legal framework regarding "local alcohol" is different in the six countries presented, as is its

*See, for example, the World Health Organization's Alcohol Database web site, on which published material on local beverages is presented.

importance within the cultures. Poverty is a thread that runs through the country chapters as a characteristic common to many consumers of "local alcohol," and limited funds and resources make it more difficult still for governments to address the root causes of harm that may be associated with patterns of consumption of these products.

For most of the participants in the studies reported in this book, legitimately produced alcohol beverages are simply not affordable. In creating the brands that they market, both nationally and internationally, commercial beverage companies adhere to quality standards and bear many costs associated with the manufacturing and marketing processes. They also pay taxes, which are passed back to those that purchase the legal product. Commercial companies cannot compete in terms of price with the "local alcohol" discussed in this book, and some might argue that prohibitive prices as a result of high taxes might drive consumers to purchase unsafe products. This may not be perceived as a problem in some countries either by the government or by the alcohol beverage industry. However, when the level of illicit production is sufficient for government to be concerned about the loss of revenue and/or the commercial alcohol beverage industry feels its profits are significantly threatened, there may be calls for action against the moonshiners. One purpose of this book is to try to help ensure that the debate about the nature of moonshine markets is better informed.

Legitimate production of alcohol is only one part of the alcohol universe. The other categories are discussed in this book and include the consumption of (1) smuggled, good-quality, legitimate alcohol; (2) good-quality, illegal alcohol; (3) bad-quality, illegal alcohol for resale; and (4) home production intended for personal/family/friends. The challenge for the public health community, government, and the alcohol beverage industry is to improve our understanding of patterns of consumption of all these products in a way that allows policymakers to focus limited resources where they are most needed. *Moonshine Markets* is a first step in that direction.

Marcus Grant
President
International Center for Alcohol Policies

Eleni Houghton
Director of Social Policies
International Center for Alcohol Policies

Acknowledgments

We wish to express our gratitude to Marcus Grant and the International Center for Alcohol Policies for making this book possible. We both believe the consumption of moonshine to be a neglected area of study and are grateful to all the authors for their important contributions to this investigation and their patience in responding to many editorial queries and comments. We are also indebted to David Thompson for his meticulous technical editing skills, and to Anthony Pyle and Andrew Wackerfuss, who prepared the manuscript for the publisher. We are especially grateful to Eleni Houghton, who coordinated the work, coped with communication problems (even with the help of the Internet), and was ever ready with suggestions as to how diverse approaches to the study could be amalgamated into a coherent whole.

Chapter 1

Introduction

Alan Haworth and Ronald Simpson

The term "moonshine" usually refers to a distilled alcohol beverage, produced and drunk clandestinely, and believed both to have powerful intoxicant effects and to be frequently dangerous. In some countries with strict alcohol control policies, moonshine may replace commercial beverages when controls become too stringent; the most extreme case of this effect is found when "prohibition" is imposed. However, there are many other reasons for the consumption of moonshine in different regions and countries, and there are often parallel markets for commercial and moonshine beverages. When this occurs, production in one of the markets goes unrecorded and the authorities may have no idea of the extent of consumption of beverages outside the commercial sector. Many reports from Africa, for example, have suggested that the drinking of "home-brew" is much more prevalent than the consumption of factory-produced beverages.

At a recent meeting in Washington, DC, it was proposed that a prospective diary-based study of "illicit" beverages be initiated in several countries in Africa. The initial brief concept paper noted that the consumption of home-brewed beers, some fruit-based beverages, palm wine, and home-distilled beverages was widespread and that much of this drinking might be harmful. There was a manifest need to study the amount of consumption of such beverages for other reasons also: Without quality control of the product, people might be consuming beverages that were dangerously contaminated; there might be dangerously high levels of alcohol and hence in the beverages, so in the absence of guidance as to alcohol content some drinkers might be at special risk of harmful drinking; without any established method of monitoring changes and trends in consumption patterns, their consumption—which might increase dramatically—could pose major public health problems; and where governments relied upon taxation of alcohol beverages, the authorities might be missing an important source of revenue.

This book describes how an idea became a reality that was taken up in a number of countries in different parts of the world, not just in neighboring countries in Africa. The need for such studies is demonstrated by the paucity of information on the drinking of these difficult-to-define beverages. There appears to have been only one major systematic series of studies (Österberg, 1989), in which four separate beverages, including "moonshine," were considered.

In the next section of this chapter, we tackle the complex question of definition and present an expansion of the original concept paper. We then examine a concept that, although not new, has increasingly come to be recognized: that in order to understand how alcohol affects humankind, one must not just develop and study crude indices of consumption but also look in detail at patterns of consumption within local social and cultural contexts. Following a commentary on alcohol policy and moonshine, including the involvement of the alcohol beverage industry, we conclude with a country-by-country review and summary.

DEFINING MOONSHINE

We began this chapter with what might be termed a "popular" definition of moonshine. Even when dealing with industry-produced commercial beverages, the terms used sometimes overlap. Most can be divided into beers, lagers, ales, and so forth (made from grain), either clear or unfiltered (opaque) and usually with an alcohol content of up to about 7% by volume; wines and similar beverages, made from fruit, with an alcohol content that is somewhat higher; and spirits, which are distilled beverages and have a much higher alcohol content than can be achieved by fermentation alone. Confusion can arise when the same beverage may be called both "banana wine" and "banana beer," for example. If the emphasis is put upon the fact that the beverage is illegal, then some forms of beer and wine need to be included. We do so in this volume. We had some trouble in choosing a name for this book precisely because the concept encompassed by our subject matter is not easily definable. An initial title, arising from a concern over increasing consumption of noncommercial beverages in some developing countries (and especially in Africa), seemed not to fit every case. We examine some of the assertions made about moonshine that are also the focus of the country reports made in this volume, noting in passing that there is again considerable overlap as one attempts to examine categories, such as between legal and commercial aspects.

Commercial or Noncommercial?

When we use the word "commercial" we tend to think of large-scale production, yet beer brewed by village women is definitely a commodity produced for sale; for example, it was said of many Zambian leaders when the country gained independence in 1964 that they would never have gone to school if their mothers had not been beer brewers. According to the present law in Zambia, such production for sale is illegal (although production for consumption at home is permitted); however, the terms "illicit" and "illegal" do not necessarily fit the case, and Zambia is not alone in that regard.

In some countries homemade beer may be the most widely consumed of all beverages. Within the village setting the distinction between beer exchanged for services and that produced for sale breaks down. Because the quantities produced may be large and therefore significant in economic terms, it was proposed that their consumption might be referred to as "unrecorded." This topic is explored in chapter 10 of this volume by Eric Single, who discusses various other designations for these beverages as well as the economic implications. The use of the term "unrecorded," while taking into account much of the consumption, is not appropriate when used generically since it encompasses neither all forms of these beverages nor the many circumstances in which they are produced and drunk. One of the problems with regard to the use of "unrecorded" consumption affects developing countries perhaps less than smaller European nations. We refer here to beverages that are legally imported into a country by travelers (and the total amount may be significant) and beverages that are smuggled into the country (Lemmens, 2000; Nordlund & Österberg, 2000; Rossou, 2000).

Cost to the Consumer

Under this heading we look not only at monetary costs but also at costs to individuals and communities in terms of behaviors related to drinking. The question of price is important to many purchasers. The beverages in question can be produced at very low cost. On the other hand, "traditional beer" made by a woman in an African village can demand a great deal of labor when water has to be collected from a distant river and firewood also needs gathering, to say nothing of the long process of growing the grain used as an ingredient; these are costs that are rarely taken into account and do not invalidate the alternative way of looking at production. For most nonfactory production there are few overheads. No advertising is needed; there is no brand name to sustain. Purchasers may be requested to bring their own containers, or the beverage maybe put in discarded bottles that had been used for other purposes. There is no tax to pay, although there may be enforcement officers (if such exist) to bribe. Often the bribe is in the form of the beverage itself. Neither are there transportation or storage costs since the beverages either have a short natural life (they go sour quickly and are drunk soon after production) or the vendor produces only those quantities that she knows, from experience, are likely to be sold quickly. Thus the low costs are passed on to the purchasers. It is sometimes the case that the private houses where many such beverages can be bought may also be the venue for prostitutes (or also sell other substances, including illicit drugs). This is often alleged, especially by those who talk of "dens of sin," but there is much variability. Studies in some southern African countries have shown that the shebeens (the name given to these formerly unlicensed drinking places) are actually associated with fewer problems related to drinking than many official taverns (Haworth, Mwanalushi, & Todd, 1981). This is understandable in that they are often small, have a regular clientele—often of close friends—and the proprietors do not wish to attract the attention of the authorities. In some cases, as shown in chapter 6, the clientele is older and less likely to be involved in harmful behavior, for example, in acts of violence. Although it is alleged that one contributory factor in the spread of HIV is drinking generally (and by implication excessive and illicit drinking), the relationship between alcohol and transmission of HIV is far more complicated (Weatherburn, 1992; Isaki & Kresina, 2000). The question of cost also comes into the drinking of some preparations containing alcohol not meant for consumption, say by "skid row" alcoholics. Some members of society may choose to drink in this way, but one cannot generalize either their style of drinking or the possible toxicity of what they have been drinking to nonbranded beverages in general.

Legal or Illegal, or Illicit?

There is often some confusion between the words *illegal* and *illicit*; both may refer to actions forbidden by law, but illicit also refers to those that may be forbidden by custom. But even when a homemade beverage is produced legally, the question arises as to whether it is being legally consumed. For example, laws may exist regarding the legal age for allowing a child or young person to drink, but these are unlikely to be enforced in remote villages, let alone homes. The consumption of beverages by those who are under age, or, in some jurisdictions by those who are already drunk, is an offense. Is the young drinker acting illegally or his father who allows him to drink performing an illegal (but not illicit) act, or the vendor an illegal act, when as part of the process of initiation into adult life the village elders begin to allow young people to have a share of communal beer?

The case studies presented in this book show that in some countries laws exist that ostensibly make the distinctions between beverages produced in officially licensed premises, under hygienic conditions, and with full quality control, and other forms of production. Laws also exist regarding the licensing of outlets and those who may imbibe. Yet they are neither fully understood in their application, nor enforced with any rigor. In some cases laws concerning the production and sale of alcohol beverages are limited in their application and provide limited or no coverage of situations relating to sale and distribution. It then becomes difficult even to define the difference between large-scale commercial production and widespread, small-scale home production of the same beverage for sale in a neighborhood. This is the situation found in Brazil with regard to *cachaça*. In Mexico as in Brazil, because of gaps in the legal framework (as Rosovsky indicates in chapter 7), there is a lack of clarity regarding the legal status of some beverages. This is overcome to some extent by use of the terms "formal" and "informal" markets and by the term "beverage of dubious origin." The term "informal" has a number of synonyms, such as "underground," "parallel," or "nonofficial," that imply illegality, although Rosovsky draws attention to the argument that the distinction should be based on the manner of production and distribution of a product and not on the product itself. The term "dubious origin" reflects this distinction, and thus, although it may include the type of beverage discussed in this book, it also includes beverages sold in hotels, restaurants, and other drinking establishments. In other words, these terms tend to add to the confusion rather than to provide a clear terminology on which policy can be based. Although these products of small-scale enterprises are without doubt produced for purposes of commerce, there is rarely if ever any question of formal quality control or of tax revenues being levied. Small-scale commerce of this kind is an important source of revenue in many countries with little or no industrialization; sometimes called the "informal employment sector," it is an important means of enabling the very poorest in society to overcome stifling poverty.

Is Moonshine Dangerous?

At the popular level, in developed countries, homemade wines are associated with village bazaars and strange ingredients, to be contrasted with expensive commercial wines of famous vintage. They are something to be accepted politely but not to be taken seriously. Sometimes a brewer of a traditional ale gains some fame, or notoriety, but if any quantity is produced for sale, the beverage soon leaves the ranks of home-brews. Indeed, many of the famous brands now sold worldwide were first produced as someone's special home-brew or spirit. It is generally believed, however, that the makers of such beverages may well produce what to the connoisseur is a barely potable drink, but not a dangerous one. At the other end of the scale, there are the beverages drunk by "skid row drunks" or "winos"—pejorative terms that condemn the beverages (which in fact are often not made for consumption) as well as the drinkers. Because the drinkers cannot afford expensive commercial products, it is believed that they poison themselves not only with ethanol but also with methyl alcohol and other toxic substances. It is possible that because "methylated spirit," often distributed with a characteristic color, was so often used by winos, the general public has come to believe that methanol is especially toxic, when in fact its toxic effects are dose related and the actual amount required to produce serious morbidity is considerable—indeed, similar to the amounts needed for ethanol to be dangerous.

Belief in the risks of drinking illegally produced beverages is very prevalent in many countries where there is widespread nonfactory production, such as in the African continent, South America, and some Asian countries. Some church and community leaders are vociferous in their condemnation, and many have no doubt that the beverages are dangerous to both physical and mental health as well as being corrupters of youthful morals. This may reflect a view of any kind of alcohol beverage, the production of which is seen as morally reprehensible (Bachiocchi, 1989). Some churches, for example, impose an absolute ban on any form of drinking, with expulsion from the congregation for a period, if not from the church altogether, being a common penalty. But more often the homemade beverage receives the greatest condemnation. Especially when beverages are described as "illicit," and usually with reference to distilled spirits, there is a popular view that the beverages are dangerous, not because of the ethanol they contain—except that they are usually seen as extra "strong" by those wishing to purchase—but because of other toxic substances. From the public health point of view, any beverage may be looked on as potentially dangerous if the amount of ethanol it contains is unduly or unexpectedly high. This may be the case for many non-factory-distilled beverages where the alcohol content has never been assessed and there is no quality control during production. The beverage might also contain other dangerous substances, including methanol and also bacterial or fungal contaminants, as well as substances deliberately added to the beverage to increase its potency. Leaders in communities may well decry the consumption of such beverages; in pointing to their dangers, they can often call upon media reports of people poisoned and even dying after drinking a particular product, on a particular occasion. Sometimes a major tragedy is reported in which scores of people may have been poisoned. Much reliance is placed on press reports and reports from the police or other officials. The following examples have appeared recently:

- Egypt: Five Sudanese men died after consuming homemade alcohol ("Five Sudanese," 1997).
- India: Consumption of illicit liquor in the Indian State of Maharashtra has claimed 20 lives and at least 49 are hospitalized because of alcohol poisoning. Toxic liquor was contaminated with methanol ("India bootleg," 1997).
- El Salvador: Local alcohol is banned by the government after 111 die ("El Salvador bans," 2000).
- Russia: Health ministry claims that 30,000 deaths resulted from alcohol poisoning because of drinking contaminated moonshine (*samogon*) or vodka (Dixon, 2001).
- Estonia: Alcohol poisoning killed 67 people. Illicit alcohol laced with methanol was said to be the cause ("Suspects charged," 2001).

- Kenya: In the spring of 2000 it was reported that 150 people died and 500 others were hospitalized after drinking hooch (*kumi kumi*) sold in two slums of Nairobi (Maharaj, 2002).
- Indonesia: At least 31 people were killed by drinking home-brew locally distilled alcohol ("At least," 2001).
- Russia: Russia's top public health official said that 47,000 people died last year of alcohol poisoning after buying tainted drinks ("In Russia," 2002).
- Uganda: "Science" gin claims nine lives in Kamuli ("'Science' gin," 2002).

In Yaoundé, Cameroon, it was reported that members of the underprivileged population are increasingly turning to potentially lethal illicit alcohol brews because they cannot afford conventional ones. According to a local police chief, "It costs less and quickly gets you high" ("Illicit brews," 2002, para. 2). In November 1997 about 20 people died in the capital from drinking "Odontol," a locally produced gin that is popular among the poor. Following this tragedy, the government attempted to prohibit the distilling and consumption of this liquor; however, the operations simply moved underground and became even more popular. Many people rely on selling Odontol and other illicit brews in order to survive and support their families. Such sales were punishable with a prison sentence, but now the public authorities are edging toward complacence. Similar stories come from Papua New Guinea, where the *PNG Post Courie*r reports how poverty drives the production and sale of moonshine in the slums surrounding Port Moresby (Sela, 2002), while in the Madang Province liquor outlets were forbidden to sell alcohol from Thursday to Sunday to reduce alcohol-induced violence (Maltrom, 2003).

It is not usually recognized that these are also often exceptional occasions. A particular batch of the beverage, produced for a particular day, may have proved to be highly poisonous; no mention is made that many other similar parties, weddings, or even funerals (because in some cultures drink is offered to the mourners) will also have been accompanied by drinking of the same type of beverage with no dire consequences. The impression given is a distorted one in that accidental poisonings are the exceptions rather than the norm. Global condemnation is easy but finding out the real facts is more difficult: It is likely that the cause of poisoning in most of these instances was never ascertained, and the media are interested in reporting dramatic events such as sudden death, not the results of chemical analyses. Since it is possible, however, that many such beverages may contain actually or potentially dangerous ingredients, we have asked contributors to this volume to have beverages commonly used in their countries analyzed to check for the presence of some of the more common toxic substances. (See chapter 11.)

Reference is sometimes made to quality control in comparing industrial with home made beverages. It should be pointed out that what might be termed informal quality control is often excised by the maker and vendor of many home products. The housewife taking cakes to a village sale will exercise informal quality control that is quite sufficient for the occasion. Likewise at the house-sale level, if an eatable or drinkable commodity finds favor with customers and appears to do no harm and the local producer uses some simple system to assess whether his or her products are up to standard, it is possible to assert that a form of quality control is in operation.

Other Terms Used in Relation to Moonshine

Other terms are sometimes also used in discussions of moonshine. A distinction may be made between "traditional" and "nontraditional" beverages—the former referring to beverages that have been widely consumed for many generations and are often home products, whereas the latter refers to commercial beverages. Yet in Mexico and Brazil, one can refer to both industrial and nonindustrial beverages, which may or may not also be traditional—for example, *cachaça* in Brazil and *tequila* in Mexico. Another term sometimes used is "local," but it lacks specificity and should only be applied when the beverages referred to have been defined more fully. For example, although they are less common than in the past, some countries have small-scale commercial breweries, fully certified as meeting the necessary hygienic and other standards set by licensing authorities, which nevertheless sell to a very small, localized market, often through outlets owned by the company. Such "local" beverages are obviously not moonshine.

It will of course be recognized by now that whatever term is used, there is none that fits the case adequately. We therefore make no apologies for the title we have chosen for the book.

PATTERNS OF DRINKING

The accounts of the various country studies set out in chapters 3–8 of this book demonstrate the many ways in which patterns of drinking may be described. Although it might have been possible in the past to describe whole countries or regions in terms of the known consumption of well-known and branded beverages, the chapters from India and from Mexico and Brazil demonstrate just how wide the variations may be. There is another sense in which "pattern" is important. There are many theories as to why people drink, but this particular concept is perhaps only now coming to be recognized and used systematically as a basis for investigation. Its importance is that drinking is examined as an activity, which interacts with many other aspects of life, at a variety of levels ranging from the biological to the social. But it has to be appreciated that the patterns to be considered can be neither purely biological nor entirely social. One cannot, for example, describe the drinking of an individual or even a population only in terms of the amount of alcohol consumed or in terms of how much subjective drunkenness occurs. Thus, one should not accept simplistic direct associations of the type that correlate quantity per capita consumption with the occurrence of problems related to drinking; there are many other moderating factors that need to be taken into account. The Lederman hypothesis (Edwards et al., 1994) and the whole concept of global correlates of hazardous or harmful drinking might well prove to be even less applicable in many countries when the consumption of the beverages considered in this book is taken into account, particularly when key bio-psycho-social aspects are also brought into the picture.

The topic is large, as is demonstrated by Heath (1998) on cultural variations, Arria and Gossop (1998) on health issues, and Acuda and Alexander (1998) on individual characteristics. The method of study—collecting information by the use of a diary—described in this volume lends itself particularly to the gathering of social data; in order not to overburden researchers, some self-reported psychological parameters of behavior were omitted. We draw attention here only to some illustrative questions that arise and appear to merit further examination. As to where people commonly drink and with whom, marked contrasts are seen both between countries and within them (e.g., India).

We have briefly discussed the considerable confusion that arises with regard to the legal status of many beverages. We can now begin to ask to what extent this influences the circumstances in which they are consumed—whether openly or hidden from the eyes of prying authorities—and whether this then allows or promotes other behaviors. Most drinking in the Russian samples took place with either family or friends and, particularly in the rural sample, at home. In Zambia, on the other hand, men do not often drink at home or with family members and are more likely to spend considerable periods with regular drinking companions.

A person's religion may have a marked influence on his or her drinking: In addition to Islam's prohibition of alcohol beverages, many evangelical churches also strongly promote abstinence. But this will also depend on the extent to which people engage in church activities, as well as the kinds of friends they cultivate. There may also be a strong religious lobby exerting pressure on those promulgating regulations on drinking.

Interrelationships may be complex, as has already been suggested with regard to HIV transmission. This provides a particularly good example of the kinds of questions that need to be asked in studying relationships in more depth. For example, do men or women deliberately seek sexual partners in certain places or is this more an accident of drinking style (e.g., more often outside the home, or without the spouse)? What of the influence of alcohol upon the individual—is it more dangerous to be slightly than very intoxicated, with regard both to agreeing to have unprotected sex and to the ability to use condoms when the intention exists? The effects may of course differ between the two sexes: It may well be more dangerous for a woman to be more severely intoxicated. Questions of this kind are further examined by Linda Bennett in chapter 9. Although outside of the psychosocial domain, the health and nutritional status of the individual affected by alcohol make up another parameter that has to be taken into account in examining HIV transmission.

ALCOHOL POLICY AND MOONSHINE: INVOLVING THE ALCOHOL INDUSTRY

In 1997 a set of principles was promulgated in Dublin concerning cooperation between the alcohol industry, governments, scientific researchers, and the public health community (Grant & Litvak, 1998). The remarks that follow respond to the question whether such principles should be applied in the case of moonshine beverages, and if so whether this is likely to be feasible.

Principles I.B and I.C, which are concerned inter alia with industry self-regulation and individual responsibility and with measures to combat irresponsible drinking and inducements to such drinking, including research, education, and support programs addressing alcohol-related problems, are not problematic in this regard. Principle I.D reads as follows: "Only the legal and responsible consumption of alcohol should be promoted by the beverage alcohol industry and others involved in the production, sale, regulation, and consumption of alcohol." This presents more of a challenge. If the principle is to be comprehensively adhered to, how can it be applied in practice to moonshine? Principle I.E, which refers to the need to ensure strict control of product safety, can only be addressed indirectly—for instance, by educating the public regarding the possible enhanced dangers of drinking some types of beverage, or by having regular checks of beverages not subject to strict regulation in countries where they are nevertheless widely available (e.g., *cachaça* in Brazil). We have some preliminary proposals, presented on a country-by-country basis, in the next section.

Although the results presented here are only preliminary, we believe that country studies have demonstrated the potential value of this technique in gathering data, which can be of great public health significance. Several of the authors have drawn attention to the need for the enforcement of existing laws or for better laws and regulations to be enacted. Any action meant to improve the human lot should if possible be based on a real situation; too little is known about the drinking of many types of alcohol beverage, even though in some countries it forms a significant proportion of total consumption.

Is there a role for government, the public health professional, and the beverage alcohol industry to play in addressing the safety and public health issues associated with moonshine markets? The question needs hardly to be asked and is indeed central to the purpose of this book, but it is certainly more complex than appears at first sight. Karlsson and Österberg (2001) examined the strictness of alcohol policies in the 15 members of the European Union, and have shown that in the 1950s only 3 had strict alcohol control policies and 9 had few controls, whereas in the year 2000 there were no countries in the few-controls category and 6 in the strict category. However, there had also been a move from imposing controls on production and sales to imposing controls according to type of beverage. It has been noted also (Moskalewicz, 1985; 1989) that control was more often exercised by governments in countries where heavy binge drinking was prevalent (e.g., the Nordic countries and the former Soviet Union), whereas control was much more subject to commercial factors in "Mediterranean" countries in which there was a different drinking style. This must be taken into account in considering the types of data presented in this book.

It is perhaps worth noting the point that the traditional policy tools used by governments to control or influence alcohol consumption include the following: taxation; control of outlets for sale and distribution; age restrictions on purchase; production controls; quality and safety controls; and education.

REVIEW COUNTRY BY COUNTRY

In the following review we ask whether the findings noted in the country reports could be helpful in determining changes in policies of governments, public health agencies, and the alcohol industry with regard to moonshine, with a view to improving both the quality and safety of drinking.

One general observation from all the country reports is that the consumption of home-distilled, home-brewed, or any illicit product (beer, wine, spirit) is closely related to income or economic status. The low cost of the products appears to be the most common factor driving the moonshine market. Most people are not able to afford the commercially produced product even if it is the preferred one. Poverty seems to be the primary reason for choosing to drink moonshine. Poverty is also one of the reasons for producing moonshine. Locally produced alcohol products are an important source of revenue for the producers, many of whom are local women using locally available raw materials. Some of these products would probably continue to exist even if more commercial products were affordable because of their use in ceremonial activities. But the main reason moonshine exists today is that it is the least expensive source of alcohol.

Such beverages generally exist outside the traditional control of governments and the public health community. This makes controlling consumption, regulating safety, controlling product quality, and generating revenue for governments a serious problem. Because of the very low price of these products in most communities, it is also very difficult for the commercial alcohol companies to sell their products.

Russia

Russia is unique among the countries studied in this volume. Zaigraev points out (in chapter 3) that Russia has a drinking culture that includes the domination of liquor (that is, distilled spirit with a high alcohol content) over other beverages, the consumption of large amounts of liquor at a time, a disinclination to consume food along with drinking, and an initial determination to get heavily drunk. Simpura and Moskalewicz (2000) provide a good review of recent government policies and programs influencing the availability of alcohol during the period from 1980 to 2000. They describe three periods during which regulatory policy changed: old regulation, deregulation, and reregulation. They also describe how these periods influenced the changes in the production of illicit alcohol. The government's various policy attempts to reduce the availability of vodka coupled with the economic turmoil in Russia have fostered an environment where illicit production and consumption of alcohol have increased dramatically, in spite of the threat of serious legal consequences.

As noted in the media reports mentioned earlier, the Russian health ministry has reported that a large number of Russians die from alcohol poisoning each year. Some of the deaths are probably the result of excessive drinking, but some are caused by drinking contaminated products. However, Nuzhnyi, in his review (chapter 11) on the chemical composition of samples of *samogon* collected during the Russian country study, found nothing that would have caused acute poisoning.

The Russian culture encourages the consumption of large amounts of liquor, thus creating strong demand for potent alcohol products. Poor economic conditions in the country, resulting in limited financial resources to buy commercial liquor, are accompanied by the availability of cheap raw materials and labor for the production of *samogon* (moonshine), high cost of commercial products, and a fear that some of the commercial vodka may be tainted.

The problem of illicit alcohol in Russia is a significant opportunity to evaluate past efforts to control the illicit alcohol trade. However, any attempt to reduce the amount of *samogon* being produced and consumed will require an extensive review of government programs, which have ranged from prohibition to legal penalties. The government will also need to determine whether it is trying to reduce harmful drinking, reduce the amount of dangerous products on the market, reduce the volume of moonshine production, or levy much needed revenue through taxation. The outcome must combine a rational alcohol policy for harm reduction, an economic development plan that increases the resources of people dependent on moonshine liquor production and consumption, and a government revenue plan that modifies the tax and licensing system to encourage the production of high-quality commercial liquor at reasonable prices.

Zambia

Haworth (see chapter 4) has presented a detailed discussion of the drinking environment and the types of beverages available, both licit and illicit, in Zambia. Commercial opaque beer is so inexpensive that there is little reason to drink the illicit beverages. However, the lowest income groups at several of the study sites appear predominantly to consume the illicit products. There is an important dynamic associated with the illicit beverages. The production of the drinks represents a significant source of income for the producer and provides an inexpensive and acceptable product to the local consumer. There is no evidence of significant safety issues associated with drinking the home-brewed beverages.

The consumption of illicit beverages in Zambia is not only the product of poverty. Many young persons in urban situations take these beverages as part of youth culture, avoiding the supervision of older persons, and this may lead more easily to hazardous drinking. But with the apparent availability of commercial beverages at very reasonable prices, it is unlikely that any government alcohol policy change would influence the production and consumption of home-brewed or distilled products. Economic development and other public health issues would appear to be of much more importance than attempting to control illicit alcohol production and consumption.

United Republic of Tanzania

The data collected by Kilonzo and colleagues in Tanzania (see chapter 5) relate to alcohol consumption in certain urban or periurban communities that may not be representative of the population in rural areas. Although it was reported that consumption of alcohol was influenced primarily by cultural practices, patterns of alcohol consumption in Tanzania appear to be changing under the influence of rapid urbanization and the breakdown of the traditional social fabric. But the high abstention rate in the study areas may be atypical, and hence there is an urgent need to survey the extent of use of alcohol (of whatever type) in the rural population before assessing what changes in policy are indicated. The primary reason for consuming local homemade products is cost. The production of these home brews is also a means of providing income to the producing families. Although drinking patterns in Tanzania appear to be changing, it does not seem that an effort to change government alcohol policy to modify the consumption of home-produced alcohol would have a significant impact on the health or safety of the country's people.

Brazil

The two most common beverages consumed in Brazil are beer (commercial) and homemade *cachaça* (see chapter 6). There is currently a very high tax on registered *cachaça* and so there is a significant incentive for producers not to be registered. Most commercial outlets selling alcohol drinks do not pay taxes. It appears that the ease of production, the availability of inexpensive raw materials, and the lack of government control provide a significant business opportunity to sell a large volume of cheap *cachaça*. If all the beer that is sold and consumed is commercial and taxed and yet is still affordable (it is the most widely consumed beverage alcohol) to low-income consumers, it is interesting to speculate about what revenue could be gained from the *cachaça* market. Because some 900 million liters of *cachaça* are sold and consumed each year in Brazil, there would seem to be an opportunity for the government, public health authorities, and the beverage alcohol industry to work toward a mutually beneficial system of control. No reports have been received of serious illness or death from drinking *cachaça*, but this does not mean that there could not be some health hazards because the ethanol content is high and dependence or other long-term ill effects could be substantial. Collaboration between *cachaça* producers and the authorities could increase government revenue and enhance the development of a thriving commercial *cachaça* industry. This effort would improve governmental control over the product's quality and safety and might allow for better information about drinking behavior for application in public health programs.

Mexico

The World Health Organization reports that in Mexico the highest proportion of the alcohol produced (73%) refers to international brand beer, followed by spirits (23%), with small-scale production of table wines (1%). Rosovsky in chapter 7 reports that among the rural population in her study, commercial beer and *pulque* were almost equally preferred, whereas among the urban study group only 3% of the alcohol consumed was homemade *pulque*. The report does not state how extensive illicit beverage alcohol is, apart from *pulque*. It is unclear what alcohol policy initiatives would be useful in addressing the issue of illicit alcohol in Mexico, or even whether it should be considered an issue, as there does not appear to be a strong public health reason. However, there may be other alcohol-related issues that require serious attention, such as those discussed by Rosovsky.

India

India presents a special challenge in terms of alcohol policies because of its immense size and diversity. In both rural and urban settings in India, no significant normative patterns of drinking have yet emerged that could be considered valid at a national level. The lack of uniformity in patterns of drinking is reflected in the range of levels of acceptability of alcohol in society. There are regions of the country where the government has instituted prohibition, such as the western state of Gujarat, and other regions such as the state of Rajasthan where the government has reportedly recommended that "heritage liquor" be promoted and marketed in order to boost tourism. A WHO report (Saxena, 1999) describes illicit liquor in India as mostly produced clandestinely in small production units, with raw materials similar to those used for country liquor (see chapter 8 for definitions of these terms) but no legal quality control checks. The alcohol content of illicit alcohol varies up to 605%. Adulteration is quite frequent, with industrial methylated spirit being a common adulterant, and occasionally it causes incidents of mass poisoning in which consumers lose their lives or suffer irreversible damage to their eyes. Cheaper than licensed country liquor, illicit liquor is popular among the poorer sections of the population. In many parts of India, illicit production of liquor and its marketing is virtually a cottage industry, each village having one or two units operating illegally.

In most regions of India the cost differential between commercial and illicit liquor is significant, primarily because of taxes. However, the cost differential between commercial and illicit liquor in Goa is quite small because of low taxes, so that the consumption of commercial products can be an equal choice alongside illicit products. It would not be possible to recommend an overall detailed policy for the whole of India. Each state needs to examine the importance of alcohol consumption with regard to both its economic and its public health aspects. Although an incident of mass poisoning may be a tragic event, long remembered by a small community, the overall incidence of alcohol-related problems may be so small, in comparison with major health hazards such as malaria, that legislators and officials may be unwilling to invest time and money in imposing extra controls on illicit alcohol use. Much will depend on the particular style of drinking found to exist as well as pressures from religious and other groups. It is possible that a tax policy that encourages the commercial liquor industry to produce high-quality, safe products at a price that is comparable to that of illicit liquor will increase government revenue and improve public health.

There are thus some common concerns about moonshine markets in all of the countries studied in this book, affecting governments, public health officials, and the beverage alcohol industry. The governments are concerned about the proliferation of illegal products and the real or potential problems of criminal activity associated with the sale and distribution of illicit alcohol. They are also concerned about the safety of consumers who may purchase contaminated products. Finally, they are particularly concerned about the loss of potential tax revenue, which in countries such as Brazil and Russia could be substantial. Public health officials are concerned about product safety and the health consequences of drinking contaminated beverages. They are equally concerned about the amount of unrecorded alcohol consumed, which might lead to problems of abuse and addiction. The beverage alcohol industry is concerned about moonshine markets because of lost opportunities for its business. In addition, it is particularly concerned about the possibility that contaminated products or especially counterfeited products may damage the reputation of legitimate beverages and lead to restrictive government policies.

SUMMARY

The reports in this volume clearly show that there is a wide variety of moonshine markets. Each one is unique and may require an individualized plan to address the issues in the country concerned. There is no universal solution. There are numerous obstacles to achieving a comprehensive modern regulation of the supply of beverage alcohol. With regard to the individual consumer, two factors that appear to drive the moonshine markets in each of the countries discussed in the book need to be taken into account. The first is the continuation of extensive poverty, which makes cheap illicit beverages more attractive than the more expensive legal beverages. The second is the easy access to cheap raw materials for home production of alcohol. Home production is often an important source of income to the impoverished producer.

Simpura and Moskalewicz (2000) suggested that to prevent large-scale moonshining the pricing policy for legal beverage alcohol has to be sensible and the accessibility of alcohol has to be satisfactory. In addition, the price of popular beverages has to be kept relatively low to maintain wide support from the working-class or unemployed people who constitute the largest group of consumers.

It is unlikely that the traditional alcohol policy tools mentioned earlier can effectively address the problems that seem to stimulate groups of underemployed and impoverished people to produce and consume moonshine. It is important that governments, public health authorities, and the beverage alcohol industry work together to provide solutions that reduce the harm associated with contaminated alcohol products and excessive alcohol consumption. This will require imaginative economic development programs backed by creative tax and regulatory systems that better utilize the available raw materials and human resources.

REFERENCES

Acuda, W., & Alexander, B. (1998). Individual characteristics and drinking patterns. In M. Grant & J. Litvak (Eds.), *Drinking patterns and their consequences* (pp. 43–62). Washington, DC: Taylor & Francis.

Arria, A. M., & Gossop, M. (1998). Health issues and drinking patterns. In M. Grant & J. Litvak (Eds.), *Drinking patterns and their consequences* (pp. 63–87). Washington, DC: Taylor & Francis.

At least 31 killed by home-brew in North Sulawesi. (2001, November 26). Agence France Presse. From LexisNexis Academic database.

Bachiocchi, S. (1989). *Wine in the Bible: A biblical study of the use of alcoholic beverages.* Berrien Springs, MI: Bible Perspectives.

Dixon, R. (2001, April 8). The vat, the bucket and some glasses—A still life of rural Russia; Alcohol: Despite repeated crackdowns, samogon is the common currency in hardscrabble villages. *Los Angeles Times*, p. A3. From LA Times Archives at http://pqasb.pqarchiver.com/latimes

Edwards, G., Anderson, R., Babor, T. F., Casswell, S., Ferrence, R., Giesbrecht, N., Godfrey, C., Holder, H. D., Lemmons, P., Makela, K., Midanik, L. T., Norstrom, T., Österberg, E., Romelsjo, A., Room, R., Simpara, J., & Skog, O. J. (1994). *Alcohol policy and the public good.* Oxford: Oxford University Press.

El Salvador bans local liquor after 111 deaths. (2000, October 12). Reuters News Service. From Factiva database—Dow Jones & Reuters.

Five Sudanese die of alcohol poisoning in Egypt. (1997, June 28). Reuters News Service. From Factiva database—Dow Jones & Reuters.

Grant, M., & Litvak, J. (Eds.). (1998). *Drinking patterns and their consequences.* Washington, DC: Taylor & Francis.

Haworth, A., Mwanalushi, L., & Todd, D. (1981*). Community response to alcohol-related problems in Zambia* (Community health research reports 1–7). Lusaka: Community Health Research Unit, Institute for African Studies, University of Zambia.

Heath, D. B. (1998). Cultural variation among drinking patterns. In M. Grant & J. Litvak (Eds.), *Drinking patterns and their consequences* (pp. 103–125). Washington, DC: Taylor & Francis.

Illicit brews flood Cameroonian market. (2002, September 10). Panafrican News Agency (PANA) Daily Newswire. From Factiva database—Dow Jones & Reuters.

India bootleg toll rises to 20. (1997, November 16). Reuters News Service. From Factiva database—Dow Jones & Reuters.

In Russia, 47,000 people died last year from bad alcohol. (2002, March 17). Associated Press Worldstream. From LexisNexis Academic database.

Isaki, L., & Kresina, T. T. (2000). Directions for biomedical research in alcohol and HIV: Where are we now and where can we go. *Aids Research and Human Retroviruses, 13*, 1197–1207.

Karlsson, T., & Österberg, R. (2001). A scale of formal alcohol control policy in 15 European countries. *Nordic Studies in Alcohol and Drugs, Eng. Supl, 18*, 117–131.

Lemmens, P. H. (2000). Unrecorded alcohol consumption in the Netherlands: Legal, semilegal and illegal production and trade in alcoholic beverages. *Contemporary Drug Problems, 27*, 301–314.

Maharaj, D. (2002, March 21). Reclaiming husbands in Kenya; Posses of women are taking aim at moonshine, which they say is wrecking families. Officials agree with them. *Los Angeles Times*, p. A1. From LA Times Archives at http://pqasb.pqarchiver.com/latimes

Maltrom, J. (2003, January 9). Madang province lacks law to control home-brew say police. *Papua New Guinea Post Courier*, p. 4. From Factiva database—Dow Jones & Reuters.

Moskalewicz, J. (1985). Monopolization of the alcohol arena by the state. *Contemporary Drug Problems, 12*, 117–128.

Moskalewicz, J. (1989). Poland. In T. Kortteinen (Ed.), *State monopolies and alcohol prevention: Report and working papers of a collaborative international study* (pp. 221–255). Helsinki: Social Research Institute of Alcohol Studies.

Nordlund, S., & Österberg, R. (2000). Unrecorded alcohol consumption: Its economics and its effects on alcohol control in the Nordic countries. *Addiction, 95*, 5551–5564.

Österberg, E. (1989). Use of home-made alcohol in Finland, 1972–1989. *Alkoholpolitikka, 54,* 199–205.

Rossou, I. (2000). Validity of political arguments in the Norwegian alcohol debate: Asso-

ciations between availability of liquor and consumption of illicit spirit. *Contemporary Drug Problems*, *27*, 253–270.
Saxena, S. (1999). Country profile on alcohol in India. In L. Riley & M. Marshall (Eds.), *Alcohol and public health in 8 developing countries* (pp. 37–60). Geneva: World Health Organization.
"Science" gin claims nine lives in Kamuli. (2002, April 22). Africa News Service, Inc. From LexisNexis Academic database.
Sela, R. (2002, October 8). Young mums sell home brew liquor. *Papua New Guinea Post Courier*, p. 3. From Factiva database—Dow Jones & Reuters.
Simpura, J., & Moskalewicz, J. (2000). Alcohol policy in transitional Russia. *Journal of Substance Use*, *5*(1), 39–46.
Suspects charged in Estonia alcohol poisonings; Death toll now at 67. (2001, October 15). Associated Press Worldstream. From LexisNexis Academic database.
Weatherburn, P. (1992). Project SIGMA Alcohol use and unsafe sexual behaviour: Any connection? In P. Aggleton, P. Davies, & G. Hart (Eds.), *AIDS: Rights, risk and reason* (pp. 119–132). London: Falmer Press.

Chapter 2

Methodology

Alan Haworth

This chapter discusses the use of a methodology that must satisfy two tests: It must be applicable in developing countries, such as those in sub-Saharan Africa, and it must address relevant issues of individual and community patterns of drinking described in the introductory chapter of this book. The study's techniques were only outlined initially in a brief conceptual paper, and participating researchers were under no pressure to adopt them fully.

The method of study proposed at the outset was the use of a diary by a relatively small number of participants. Because the study would be essentially descriptive (although also allowing for limited numerical analysis) and focusing on usually unrecorded consumption, it was suggested that data be collected in areas where drinkers of such beverages were most likely to be found. Only adults (those over the age of 15) would be invited to participate. It was recommended that the basis of sampling should be the family, and preliminary data on family structure also needed to be collected. Consumption was to be recorded on a daily basis in a diary, which would have at the front a list of open-ended questions that could be answered with respect to key elements of the previous day's drinking, namely, date and day, place where drinking took place, the types and quantity of alcohol consumed, drinking companions, whether there was anything special about the drinking occasion, and the effects of drinking, both desired and problematic. Diaries were to be kept by individuals under the guidance of research assistants who would visit or meet with the diary keeper as often as necessary to ensure accurate recording. Recording would take place over periods of 4 or 5 days, to a total of about 30 days, and these periods would be arranged to examine drinking at specific periods such as weekends, month ends, and around any public holiday or festival. The diary method presents some ethical problems in that where diaries were kept by several members of one family there would be the possibility of coercion or lack of confidentiality, and these issues would be given special attention in the training of field-workers.

In addition to describing how they have adapted the techniques, the authors of these chapters drew on other sources of information for data where they have felt that this was more appropriate given the present stage of the work. Thus, researchers have relied on statistics for the production and distribution of beverages from Mexico and Brazil, rather than on the person-centered approach implicit in the proposed method. Both techniques are essential to gaining a full understanding of the dynamics of drinking patterns, although this chapter focuses only on the techniques recommended for examining individual consumption patterns within particular contexts. Because the techniques have not been widely used especially with regard to the consumption of what are sometimes defined as illicit beverages in the countries concerned, some of the studies described in this book are tentative and essentially focus on the possibility of further exploration.

REVIEW OF ALTERNATIVE METHODS OF DATA GATHERING ON DRINKING PATTERNS

Rapid Assessment

Various methods for the study of drinking in populations have been developed, but they yield data of variable relevance and accuracy, and some demand considerable resources. In recent years, rapid assessment techniques using key informants and focus-group discussions have become popular. The schedules or menus for interviews and focus groups may include questions of a quantitative nature, but because these are expected to be viewpoints of individuals or groups, the report containing apparent quantitative data may be misleading; percentages that refer to the number of informants giving a particular estimate provide no information on the frequency or extent of an actual behavior. There is a tendency for some readers, including policymakers, to misunderstand the nature of the data and to believe that they are quantitative and reliable.

There are other disadvantages. For example, because focus-group discussions rely on fallible memory, the opinions and pressures of other group members as well as the injudicious selection of key informants with strong biases can lead to serious misinformation. A recent study from Zambia (Mukuka, 2000) reported (in a rather ambiguous statement) that 40% of 301 makers and sellers of an illicit spirit who participated in focus groups organized by an enforcement agency had "experienced" the death of an abuser of the beverage. No definition of the word "experienced" in this survey was given, but it appeared that no informant admitted to actually having seen death related to the beverage. Nevertheless, it was widely reported that the high toxicity of the beverage had been confirmed.

This is not to say that focus-group discussions are entirely inappropriate in studies of this kind. Rather, because behaviors related to drinking are being studied, focus groups will often allow exploration of the meaning of apparent or possible associations between behaviors and may well provide a possible normative basis that will require further study. For example, the tendency of Zambian men not to go out drinking with their wives may relate to their seeking other sexual partners because of the belief that sexual intercourse with their wives during the early stages of breast-feeding infants could spoil the milk. Having established this possible linkage, a technique using path analysis may then be appropriate. Focus groups may also be used in the early stages of a survey when more information is required, not just the quantity and frequency of alcohol consumption.

The introductory chapter to this volume, as well as the country accounts, draws attention to the enormous range and variety of beverages needing to be included. Against the types, and in some cases brands that are known worldwide, we must set the names of beverages known only in a particular area of a city, district, or province. When asked about their drinking, many informants may refer only to the well-known brands and omit others they consume unless the interviewers inquire about them.

Population-Based Surveys

Although preliminary rapid assessment has its limited place, the large-scale, population-based assessment is likely to be expensive. The cost greatly increases if a study is carried out in distant rural areas of a large country with a widely dispersed population. But the cost cannot be measured only in monetary terms. Smaller developing countries lack the skills to conduct such surveys, and although some individuals have the ability, a lecturer, say, in a small university department may have little time for research.

During a large-scale survey, large numbers of interviewers may be required because, in populations with lower rates of adequate literacy, self-completion questionnaires cannot be used. By "adequate literacy" we mean an ability to understand questions, instructions, and underlying concepts in the written form of the language used, as opposed to just oral usage of the more domestic form of the language. The latter type of language usage is found in some major cosmopolitan cities where a lingua franca is being formed from a fusion of those elements of other languages used most commonly in ordinary social intercourse. In such cases interviewers must be well trained and also well supervised, and this requires a high ratio of supervisors to research assistants. However, it is in organizing supervision that a problem often occurs, demonstrating that recruitment at this level may be difficult.

Many survey instruments also rely on memory for recording drinking events over long periods, but even information about the frequency of drinking during the previous month, for example (a period often recommended), can be only an estimate. A survey undertaken in Zambia in 1977 (Haworth, Mwanalushi, & Todd, 1981) showed that this limitation applies also to a period of 1 week, where the following problems were encountered:

> The totals may represent seven days' drinking—or one day—for they were calculated for every respondent who had drunk at least once during the week before the interview . . . numbers of respondents giving details of their drinking falls dramatically from the most recent occasion to those more remote in time. . . . One may well suspect that respondents were unwilling to make the effort to remember details of their drinking if asked to describe too many occasions and information becomes increasingly inaccurate. (pp. 15–16)

In this case, it seems that the length of the questionnaire and the amount of detail required created a strong inhibitory effect on detailed recall, and this resulted in the distortion demonstrated. Any method relying on recall of specific behaviors is subject to inaccuracies of recall because of simple forgetting, telescoping (distorting the recency of memorable events), and exposure to misleading information since the event. There can also be intentional overreporting and underreporting on sensitive issues.

Timeline Follow-Back

Although a specific technique called timeline follow-back, or TLFB (Sobell & Sobell, 1992, 1995), was developed to overcome this problem and has had considerable success, it requires highly skilled interviewers. Retrospective, self-reported data on daily alcohol consumption are collected, using various interview aids. For example, the respondent may write the report on a calendar marking salient events (i.e., visits, parties, football matches, etc.), identifying lengthy periods of abstinence or patterned drinking (such as drinking only on the weekend), and anchoring the highest and lowest quantities consumed in the target interval.

Although the TLFB method enables the collection of data from single interviews, the diary method demands a closer and longer association between research subject and research assistant and facilitates the collection of more accurate data on the context, frequency, and quantity of drinking. General population surveys also tend to focus on total amounts of alcohol consumed and frequency of drinking in a global manner, but are seriously limited in their exploration of the total context of drinking. Sufficient detail on the context can rarely be remembered accurately enough even for those occasions of highest self-estimated consumption. Although such data are useful in demonstrating that there is an association (although causality may still be in doubt), it is rare to examine potential mechanisms for increased risk behaviors, for example, with regard to HIV transmission, although the TLFB method has been adapted for this purpose (Carney, Tenner, Affleck, Del Boca, & Kranzler, 1998; Weinhardt, Cory, Maisto, & Gordon, 2001).

Alcohol Use Disorders Test (AUDIT)

The individual country chapters show that one particular survey technique was used for the selection of populations for further study in some countries. The Alcohol Use Disorders Test (AUDIT) was developed by the World Health Organization in 1992, and validation studies were also conducted in collaboration with the organization, which promoted its use in many countries. However, its use was not recommended for the actual survey of drinking behavior, because it is essentially a screening instrument for the detection of possible dependent or hazardous drinking, which then needs to be confirmed by follow-up clinical or other assessment methods. In addition, AUDIT relies on the recall of drinking behaviors over long periods (e.g., 1 year) and is thus subject to the problems mentioned already.

DIARY METHOD COMPARED

Where quantities of alcohol are to be recorded, all methods need an efficient scheme for reducing different types of beverage, served in varying quantities, to a standard measure. AUDIT (Babor & Grant, 1989) uses a "drink" as the unit, a drink being defined as that quantity of a beverage containing 10 grams of alcohol, plus or minus 25%. This definition takes into account most drinks and allows easy international comparisons. Sometimes other measures are used. The real problem occurs with those beverages that are drunk from a communal cup or mug, as in the case of opaque (unfiltered) beverages commonly drunk in many African countries.

It is not usually possible, in a single interview, for an informant to calculate the quantity he has consumed on multiple occasions because the process involves a number of steps and not everyone is adequately numerate. For example, it is necessary to check the number of people in a drinking group, to check the size of the drinking mug and the quantity of beverage served, and then to calculate the amount the individual might have taken after checking the number of mugs actually bought and consumed. In the case of an individual making entries in a diary, the process of calculation becomes easier with time (and help from the research assistant) and the regular drinker can, after a few days, usually enter the amount on the basis of previous experience.

Instead of asking participants to cast their mind back over periods of several days, or even weeks or months, the longest period required of a regular diary keeper should be 24 to 36 hours. Most studies using this technique have been concerned with frequency of drinking and quantities consumed. Feunekes, van de Veer, van Staveren, and Kok (2000) reviewed 33 methodological papers published after 1984 on alcohol intake assessment by five main methods: quantity frequency; extended quantity frequency; retrospective diary; prospective diary; and 24-hour recall. The mean levels of alcohol intake differed by 20% between these methods. An important finding was that methods inquiring about individual beverages were much more accurate. Also, because long-term recall is not necessary when the prospective diary method is used, recall of intake of individual beverages is much easier. Because the study proposed to explore the drinking of "nonstandard" beverages, this is relevant to our choice. Wynn (2000) compared a daily diary with timeline follow-back and noted that the diary produced significantly higher estimates of consumption for "normal" subjects.

Several studies (Carney et al., 1998; Leigh, Gilmore, & Morrison, 1998; Lemmens, Tan, & Knibbe, 1992; Shakeshaft, Bowman, & Sanson-Fisher, 1999; Whitty & Jones, 1992) showed that diary methods produced higher frequency levels but had little effect on reports on quantity. Lemmens et al. (1992), in discussing prospective diary, retrospective 7-day recall, and summary measures such as a quantity-frequency index, state that coverage was highest with a diary method, suggesting a high validity; underreporting in comparison with the diary method was higher in the frequency than the quantity domain. Shakeshaft et al. (1999) also compared two retrospective measures: a retrospective diary and a quantity-frequency index. More alcohol consumption was reported using the diary, and this method also detected a greater proportion of heavy and high drinkers.

A question that frequently arises with any survey of possible socially unacceptable or illegal behaviors concerns the truthfulness of the participants. The convergence and consistency of answers by the individual and from aggregated results usually offer reassurance, but the studies just reported indicate that, in particular, underreporting does occur. One of the most accurate means of gathering data is by being absolutely impersonal—that is having participants answer questions using a computer (Carney et al., 1998). It should soon be possible to use such methods in developing countries, but there are still constraints. The least satisfactory method would appear to be the use of a group of interviewers who have one-time sessions with time constraints, because both interviewer and informant may be tempted to cheat. The method proposed for this project involves multiple contacts between interviewer and informant and the use of counseling techniques, so that a collaborative and trusting relationship may develop.

One reason for proposing this type of relationship is the likelihood that informants may be asked to report on possibly illegal activities. When considering illicit beverages whose manufacture, sale, and consumption may result in legal sanctions, it is difficult to collect adequate data, because informants in any survey may be reluctant to put themselves at risk of arrest or punishment.

The proposed method, which involves collecting sets of data over a longer time period, reduces the need to recruit a large sample and reduces the overall logistical costs—an important consideration given the limited resources in developing countries. The diary method, which uses a relationship, bears some resemblance to classical ethnographic research. Even though various skills of the ethnographer are reduced to a set of mechanical actions (diary keepers are not invited to be expansive in their accounts), the data collected can still prove to be more rounded and context sensitive than that collected using most other methods.

Because items of information can be collected according to certain key periods such as weekends, month ends, and festive occasions, the diary provides a parsimonious method for discerning possible patterns. However, although relatively few subjects may be participating, the large amounts of data generated result in labor-intensive data entry. It may be worthwhile to improve the design of diaries so that data may be collected in machine readable fashion, allowing flexibility for the extra information that can only come with some freedom in the style of entry.

All methods face an important limitation in multiethnic societies, where it may be difficult for participants to fully understand a question. The diary method has an advantage over other methods in this regard, because while checking and discussing initial entries, the understanding of each question can be probed and any misunderstanding corrected. As with all methods, the diary method relies on the respondent's honesty, but by having regular support and supervision from a research assistant, inaccuracies should become more readily apparent and understood to be unacceptable.

There are, however, other limiting factors in the use of a diary, including the time it takes research assistants to visit each diary keeper and the need to continue with this exercise over a long period. Careful training and supervision are needed, and the supervisors must be fully aware of the purpose of each activity, each question, and the relevance of each piece of information required. It is important to have a system in which all supervisors and research assistants keep dairies of their own supervisory activities.

A case for having preliminary focus-group sessions and key-informant interviews has already been made. Training sessions with research assistants can be made into useful focus-group discussions in which their knowledge of beverages and patterns of drinking can be explored and biases in the way any behaviors are viewed can be eliminated. Training programs sometimes reveal that an individual holds such strong views about the unacceptability of an activity that they could not be permitted to continue the work. This "qualitative" work may reveal the need for menu items that may otherwise have been overlooked. For example, in the Zambia survey, it was noted that there was a change in "preferred" alcohol content of certain beverages as compared with previous reports. But the word "preferred" may not be appropriate, because the impetus for change may have come from manufacturers or vendors and not from drinkers. The change was not detected in the exploratory discussions, and on finding unexpectedly high concentrations of alcohol in samples collected at outlets, initial suspicion was of some mistake having occurred in the collection and delivery of the samples for analysis. Only when further samples were collected and analyzed, did the question of change in preference arise.

The keeping of a drinking diary should be one of the most accurate methods of studying the context of drinking, provided that participants are careful, truthful, and assiduous in their task and receive the protection of anonymity. An advantage of having combined brainstorming and training sessions for research assistants is that relatively unskilled personnel may be used, provided that they fully understand the task they oversee. Criteria for the selection of samples were to be decided in each country, but it was suggested that the sampling unit should be the household and that as many adult members as were willing should be invited to keep diaries. Research assistants were to ensure that each diary should be seen only by the diary keeper and the assistant; in particular, parents and other family members should not see the contents. Although this rule undoubtedly discouraged some participants, it helped ensure a high degree of disclosure. In general, it is recommended that where interference by a senior member of the household occurs, that household should be dropped from the survey. Kilonzo (see chapter 5) listed a number of other problems in Tanzania, including a lack of privacy in some homes, confidentiality issues, and failure to keep appointments, but concluded that research assistants and supervisors had the impression that a diary was an acceptable method and that the majority of respondents were accurate in their recording and did not resent the numerous visits to their homes.

In a study of this type it is essential to obtain samples of all beverages consumed from each of the sites and on several occasions. Quantities of beverages being served in a variety of containers also need to be checked. In the case of homemade beverages, it is also advisable to measure the ethanol content and to check for various other products of fermentation, contaminants (e.g., from the water supply or from vessels used in manufacture), and additives, meant to increase the potency of the beverage. The findings from those countries where this information was obtained are presented in chapter 11, which includes a detailed analysis of the implications.

COMMENTS ON VARIOUS COUNTRY APPROACHES

Individual investigators were invited to adapt the methodology to their circumstances and hence it is possible to draw on a wide range of experience in using this method of study. Four countries decided to adopt a variant on the method for selecting study areas used in the Zambia and to seek areas, families, or individuals with a high likelihood of drinking beverages of interest. Details are given in each chapter, and the Zambian chapter gives the rationale for seeking those who were likely to be heavier drinkers. In the case of India, data were collected in four very different regions, and this gave an opportunity for the vast differences in drinking practices in the subcontinent to be demonstrated and reviewed. Data were also collected from both rural and urban settings, but within these, members of specific groups (not comparable between sites) were invited to participate. The same policy was adopted in Russia, and data were collected from widely separated regions with somewhat different populations in terms of income and employment opportunities. The Mexican samples were sufficiently contrasted also to draw attention to some of the differences seen in drinking styles. Both studies from Africa took samples from one region only, although there was more contrast between the Tanzanian than the Zambian samples.

In India, known heavy drinkers were approached and further selection was carried out using the WHO AUDIT, described earlier. The cut off score for inclusion in the study was 8+. In the case of the Russian study, volunteer families were selected with an average income or slightly below the average for the region. All adult (over 16 years of age) members of the families were invited to participate after giving an interview in which they provided basic demographic information and responded to questions on alcohol in general and on *samogon* consumption. In Brazil, health personnel already involved in assisting those with alcohol-related problems were able to identify suitable families. Health workers also helped identify participants in both the rural and urban areas selected in Mexico. In all these cases, the focus was on examining the drinking of all types of alcohol within the context of regular consumption. Although this means that the results cannot be extrapolated to populations in general, it has allowed for the method of data collection to be tested, for results likely to reflect the drinking styles of the (drinking) population at large and to give some indication on what may need to be done in any large-scale study. In the case of Zambia, it was possible to compare the results of this method of data collection and confirm that it could provide a picture similar to that provided by a larger, more expensive, population survey.

The approach taken by the Tanzanian group was somewhat different in that more attention was given to solving problems relating to actual fieldwork. The sample was a random one of clusters of household and included many nondrinkers. Because they too were invited to keep diaries in case any should begin drinking during the period of data collection, an initial problem arose with regard to their reluctance to participate (none was reported as taking any alcohol during the project). There was some opportunity to examine such variables as socioeconomic status and drinking. The detail provided by Kilonzo is important from a practical point of view. Because the data were collected during brief periods of 4 or 5 days, it was possible to treat the first period as a pilot run (and discard the data in terms of analysis) while practical problems were being solved. It is noteworthy that all the problems mentioned appeared to have been dealt with satisfactorily, and the ensuing data collection went smoothly, confirming the practicality of the method. Kilonzo rightly stresses the need for extremely careful supervision of research assistants and the need to keep full and detailed records on the process of data collection as well as the content.

Various techniques were adopted for recording the data. In the case of India, although the approach was that of keeping a diary, data were collected by interview, covering a period of 1 week, repeated on four occasions. In one case a precoded format was used (Russia) and in others a much simpler format. No reports were received on the preferences of the participants. Data entry is always easier from a precoded format, but the method allows for the collection of so much detail that not all the data collected were used in analysis by every country. Investigators need to have a very clear appreciation of the problems of managing bulky data sets from an apparently simple method of collection.

Background demographic data were collected in all locations, but in some cases further information was obtained on the household's drinking. In Tanzania it was felt that this might have influenced the data being provided but this was noted during the first period of data collection and could be attended to. There was no mention of any payment being made in India, Russia, or Brazil, but in the other three countries it was felt that reasonable compensation should be paid for time taken and for other expenses. In the case of Mexico a payment was made per household and this may have caused fewer participants to participate per household, than if individual diarists had been compensated. Researchers in Brazil reported 10 of the 91 (11%) participants withdrawing from the project, and in Tanzania 1 withdrawal. Other studies reported none, although in Mexico there was some difficulty in recruitment, because some participants were not at home when field-workers visited and others gave various excuses; one had joined Alcoholics Anonymous.

CONCLUSION

One of the objectives of the project was to test whether using a diary method for obtaining a wide range of data on drinking, using minimal technical resources, is feasible, efficient, and yields reliable and valid data. By its very nature, the last two objectives could not be formally tested, but the literature reviewed suggests high validity; it would be necessary to use alternative methods of data collection in order to confirm that the method used produces more accurate information on frequency of drinking and amount of alcohol taken. There were no reports of difficulty in recalling necessary information. Because diary keeping is continued over a long period (i.e., as long as 2 months) with regular assistance and supervision from nonjudgmental field-workers capable of building up a relationship of trust, it is likely that high reliability can be attained.

All of the countries reported some difficulties in execution but equally were able to report rapid resolution after "teething problems" had been overcome. Minimal resources were required in the field, and even simple school exercise books could be used for the recording of data. The method is especially suitable for countries with several ethnic/linguistic groups and where considerable numbers of participants may have limited capacities to read and write. The fact that large volumes of very systematic data could be obtained from relatively small numbers of participants, with minimal expense, is also a positive feature, although the technique demands extra work in data coding and entry. The level of analysis asked for does not require highly sophisticated statistical techniques, but considering the actual volume of data, the use of a computer is obligatory; desk-top computers are now "standard equipment" in every country. It may be concluded that the use of the diary method has many advantages, especially with regard to examining the consumption of illicit beverages, and can be recommended.

REFERENCES

Babor, T. F., & Grant, M. (1989). From clinical research to secondary prevention: International collaboration in the development of the Alcohol Use Disorders Identification Test (AUDIT). *Alcohol, Health and Research World, 13*, 371–374.

Carney, M., Tenner, H., Affleck, G., Del Boca, F. K., & Kranzler, H. R. (1998). Levels and patterns of alcohol conusmption using timeline follow-back, daily diaries and real-time "electronic interviews." *Journal of Studies on Alcohol, 59*, 447–454.

Feunekes, G. I., van de Veer, P., van Staveren, W. A., & Kok, F. J. (1999). Alcohol intake assessment: The sober facts. *American Journal of Epidemiology, 151*, 105–112.

Haworth, A., Mwanalushi, M., & Todd, D. (1981). *Community response to alcohol-related problems in Zambia (Community health research reports 1–7)*. Lusaka: Community Health Research Unit, Institute for African Studies, University of Zambia.

Leigh, B. C., Gillmore, M. R., & Morrison, D. M. (1998). Comparison of diary and retrospective measures for recording alcohol consumption and sexual activity. *Clinical Epidemiology, 51*, 119–127.

Lemmens, P., Tan, E. S., & Knibbe, R. A. (1992). Measuring quantity and frequency of drinking in a general population survey: A comparison of five indices. *Studies in Alcohol, 53*, 476–486.

Mukuka, L. (2000). *A baseline study of the extent to which illicit alcohol (kachasu) is used and abused by peri-urban communities in Lusaka and Mazabuka*. Lusaka: Center for Social Policy Studies, University of Zambia.

Shakeshaft, A. P., Bowman, J. A., & Sanson-Fisher, R. W. (1999). A comparison of two retrospective measures of weekly alcohol consumption: Diary and quantity/frequency index. *Alcohol, 34,* 636–645.

Sobell, L. C., & Sobell, M. B. (1992). Timeline follow-back: A technique for assessing self-reported alcohol consumption. In R. Z. Litten & J. Allen (Eds.), *Measuring alcohol consumption: Psychosocial and biological methods* (pp. 41–72). Totowa, NJ: Humana Press.

Sobell, L. C., & Sobell, M. B. (1995). *Timeline followback user's guide*. Toronto: Alcohol Research Foundation.

Weinhardt, L. S., Cory, M. P., Maisto, S. A., & Gordon, C. M. (2001). The relation of alcohol use to HIV-risk sexual behaviour among adults with a severe and persistent mental illness. *Journal of Counseling and Clinical Psychology, 69,* 77–84.

Whitty, C., & Jones, R. L. (1992). A comparison of prospective and retrospective diary method assessing alcohol use among university undergraduates. *Public Health and Medicine, 14*, 264–270.

Wynn, P. A. (2000). The reliability of personal alcohol consumption estimates in a working population. *Occupational Medicine, 5,* 322–325.

Chapter 3

The Russian Model of Noncommercial Alcohol Consumption

Grigory Zaigraev

HISTORICAL DEVELOPMENT OF RUSSIA'S ALCOHOL CULTURE

In Russia, the consumption of noncommercial alcohol (i.e., homemade alcohol beverages) among rural populations mainly concerns *samogon*—a distilled spirit with an alcohol content similar to vodka. The distinguishing features of *samogon* consumption can be identified and correctly understood when Russia's local peculiarities and its historical patterns of alcohol use are taken into account.

Compared to many other countries, Russia developed a less disciplined form of alcohol use because of a confluence of economic, social, cultural, geographic, and climatic factors. As a result, negative consequences of alcohol consumption, including alcohol abuse and alcohol-related disease, are much more prevalent.

Typically, the Russian drinking culture is characterized by the predominance of liquors over other beverages, the consumption of large amounts of such distilled liquors on one occasion, a disinclination to consume food when drinking, an initial determination to get heavily drunk, and the existence of many drinking traditions that have turned alcohol consumption into an inalienable element of the Russian lifestyle.

State alcohol policy, which since the 15th century had steadily promoted a vodka-based pattern of consumption in order to increase revenues, has played a decisive role in forming this drinking culture. The government used harsh repressive measures to suppress home production of light alcohol beverages (mash, beer, or *medovukha*, an alcohol beverage made of honey), to establish a state monopoly on "bread wine" production, and to promote and spread "tsar's taverns"—places for the official sale and consumption of vodka only, without any food. During the 18th century, beer-producing enterprises were closed throughout Russia, except in Moscow and St. Petersburg. As a result, by the beginning of the 20th century vodka made up 93% of all beverage alcohol consumed. Subsequent large-scale state action—bans on alcohol beverage production and sale that were intended to combat drinking—resulted in even greater unruliness in Russia's alcohol consumption culture.

Samogon's Place in Russia's Alcohol Culture

Samogon's importance in modern Russian patterns of alcohol consumption resulted in part from these government efforts. The state's attempts to eliminate the home production of alcohol beverages and reduce overall consumption through periodic bans failed, in part because of conflicting interests such as the state's long-standing wine monopoly. Given a steady level of demand for alcohol among the general population, home production actually increased during these efforts. The state's alcohol policy had a very strongly negative impact on drinking culture, introducing massive *samogon* consumption as an enduring practice.

Samogon production has always been suppressed, but on a different scale and with varying degrees of stringency. *Samogon* production in Russia would decrease and then increase again, with periods of underground activity followed by rapid peaks. During the 20th century, *samogon* production peaked in Russia during three major periods of expansion: 1914–1925, 1985–1988, and 1996–2002.

Widespread *samogon* production began for the first time after the introduction of the temporary prohibition on vodka sales by the tsarist government in 1914. This measure was aimed at preventing drunken disorders that were supposedly hampering First World War mobilization efforts. However, by the second half of 1915, Russians were already flouting the ban with massive and widespread production and consumption of *samogon*, lacquer, polishing varnish, and other alcohol-containing liquids, which resulted in numerous alcohol poisonings and a rise in alcohol-related diseases.

During the revolution and civil war, the Soviet government continued to prohibit the production and sale of alcohol beverages throughout Russia. During the civil war, general anarchy and vodka shortages further encouraged *samogon* production. According to a State Statistical Committee survey, in the summer of 1923 more than 10 million rural households—10% of all households—produced and sold *samogon*. That level of production required an annual expenditure of 2.5 million tons of grain.

In response to the increasing scale of *samogon* production, the Soviet government decided in October 1925 to permit the production and sale of 40° vodka. However, *samogon* had already become entrenched among a significant part of the population, creating long-term problems even after prohibition efforts had stopped. According to data for 1927 from the Russian Federation Central Statistical Department, total per capita official alcohol consumption in rural areas (which accounted for two-thirds of the country's population) stood at 1.8 liters, compared to 7.5 liters of *samogon*. In other words, the ratio of *samogon* to vodka was more than 4:1.

During the following decades, especially during the years after the Second World War, production of *samogon* was to some extent controlled by strict laws that entailed criminal prosecution for the production and storage of homemade liquors. Punishment ranged from 1 to 3 years in prison, with the confiscation of property. Nevertheless, even such severe penalties were not sufficient to significantly diminish *samogon* production. According to the calculation of experts from the Central Statistic Department, in the early 1980s the per capita consumption of 40° *samogon* was at least 5 liters.

In 1985 the government began an ambitious antialcohol campaign, which by curbing alcohol production and sale created a second wave of *samogon* production from 1985 to 1988. During those years, official annual per capita alcohol sales decreased from 10.4 to 3.8 liters of absolute alcohol, or some 63%. But with the same level of demand among the population, the shortfall in alcohol beverages in turn caused the widespread proliferation of *samogon* production, including in regions where it had previously been limited.

In 1988, 572,000 individuals were called to administrative account for *samogon* production, 5.7 times the 1985 figure. According to police department estimates, this accounted only for about 15–20% of total *samogon* producers.

As to the volumes of *samogon* production in the same year, the USSR State Statistical Committee indicator estimated it at 8 liters or 3.2 liters of pure alcohol per capita.

During that period *samogon* consumption was observed in almost all social groups. A 1988 sociological study conducted under the supervision of the author in a number of regions revealed that 68% of agricultural and 45% of industrial workers, as well as 24% of the intelligentsia, drank homemade alcohol beverages with their closest social contacts.

During the 1990s, the fall of the Soviet government had a significant impact on the pattern, scale of production and consumption of homemade alcohol beverages and their prevalence among various population groups. In January 1992 the state ended its monopoly on alcohol production and sale, and at the same time decriminalized noncommercial home production. Those measures, combined with a dramatic fall in the standard of living, helped once again to expand the role of *samogon* as newly impoverished rural populations turned to it in order to meet the demand for beverage alcohol. A large-scale population survey conducted in 1999 by the Research Institute of the Ministry of Internal Affairs found that almost half of the respondents, and three out of four rural inhabitants, either cited massive *samogon* production in their region or noted its increase.

The experiences of professionals working on alcohol-related problems confirms these results. The majority of 130 experts interviewed by the author in different parts of Russia (Briansk region, Oriol region, Lipetsk region, Voronezh region, Tambov region, and Omsk region) considered *samogon* the main alcohol beverage for 60 to 70% of rural inhabitants.

Although this situation remains generally true of *samogon* production across Russia, it does not shed light on the patterns of *samogon* consumption. An assessment of the consumption model requires consideration of the frequency, intensiveness, conditions, and motives for this type of alcohol use both among the general population and in particular age, gender, and social groups. Account must also be taken of the changes in the scale of *samogon* consumption and the general drinking culture over time, which have significantly expanded during the last 15 to 20 years. These variables have all been included in the study described in this chapter.

STUDY PROCEDURE

In accordance with the methodology proposed by the International Center for Alcohol Policies, this comparative study examined 75 typical rural families from three typical regions of Russia: the Voronezh region, the Nizhegorod region, and the Omsk region, which are located in different geographic and economic zones. For the first time, the study obtained sociological information on rural Russian alcohol consumption, *samogon* consumption, and the place of *samogon* in everyday life.

The study proceeded in two stages. First, all adult family members (over 16 years of age) were interviewed according to a questionnaire that included questions on respondents' gender, education, social status, and attitudes to alcohol consumption, both in general and specifically regarding *samogon*. Afterward, each respondent provided information on the amount and type of alcohol beverages consumed during specified periods of time (the last 4 days of the second and fourth week of each month) during 4 months (February—May 2001), a total of 32 days of data. They completed the forms under the observation of 15 research assistants selected from a pool of medical doctors, social workers, and Ministry of Internal Affairs staff. Each research assistant worked in one community, with five families selected for the study.

The survey population was composed of 210 individuals, of whom about 20% were below 30 years of age, 34% aged 30–49 years, 22% aged 50–59 years, and more than 24% aged 60 years or older. The largest group in the sample (49.5%) were working, while 33.3% were retired and 6.7% were students.

SURVEY RESULTS

Analysis of the responses with regard to structure, frequency, volume, motivation, and settings of alcohol consumption produced the following findings.

First, *Samogon* production in rural areas has indeed increased in recent years, as has the relative consumption of *samogon* compared to other types of beverage alcohol. In all three regions, the majority of respondents (up to 80%) preferred to drink *samogon*, while 32–48% also consumed vodka along with *samogon*, and 10–15% drank wine, home-brewed beer, and other homemade beverages together with *samogon*. According to the survey, rural Russians drank 4.8 times more *samogon* than vodka—a higher ratio than in the previous highest period of *samogon* production in Russia, the 1920s, when the proportion between *samogon* and vodka had been 4.1:1. However, during that period the enormous production and consumption of *samogon* were the result of a dearth of vodka under the "dry law" of 1914–1925. The present explosive increase in *samogon* production has a very different cause, related more to economic and legislative aspects of state policy on the production and sale of alcohol beverages.

In the new survey, with plenty of vodka on sale at retail outlets, almost two-thirds of respondents explain their preference for *samogon* by the high price of official vodka, which meant that in their present economic situation they often simply did not have money to buy it. Of respondents mentioning that situation, 60–70% preferred to buy *samogon*, which had an average price of 15–20 rubles per 500 ml. bottle—half the 40–50 rubles charged for a bottle of vodka. The other 30–40% made *samogon* themselves, almost always with sugar. Taking into account the price of sugar (14–15 rubles per kilogram), the average cost of producing 0.5 liters of *samogon* was 8–10 rubles, which is one-fourth the price of officially sold vodka.

A smaller group (55% of all respondents) explained its preference for *samogon* by citing the risk of possible poisoning with counterfeit vodka—an increasingly common misfortune. (There were more than 355,000 cases of lethal alcohol poisoning in Russia in the 1990s not including *samogon*.) That group seemed confident that homemade *samogon* was of better quality than vodka and consequently less dangerous for human health, a claim that has been bolstered by laboratory testing. Only 15% of *samogon* samples tested were of low quality and high toxicity, so that there were fewer lethal alcohol poisonings from *samogon* than from low-quality vodka.

However, this generally satisfactory quality of *samogon* applies mainly to types prepared for personal consumption or for selling to close acquaintances. *Samogon* produced for sale in the general market, to unfamiliar people, is often of poor quality and diluted with various harmful components such as dimedrol in order to increase its psychoactive effect.

Second, the study found high levels of habitual drinking and the early stages of alcohol dependence among the majority of rural inhabitants surveyed, in terms of both frequency and volume of alcohol consumed. Analysis of the data on frequency, motives, and dynamics of consumption showed that alcohol use in rural areas was increasingly unstructured. Half of the respondents drank alcohol beverages at least four times a week (65% of men and 23% of women), more than one in four (27%) two or three times a week, and only one in five two to four times a month. Although more than two-thirds of the heaviest drinkers were men, women have dramatically increased their alcohol consumption: 59% of them drink alcohol at least twice a week, a dramatic change from the Russia of 50–60 years ago.

Third, alcohol consumption controlled for age distribution has displayed an unexpected picture. As can be seen from Table 3.1, individuals in older age groups drank alcohol at significantly higher rates than those in younger groups. Attitudes toward different types of beverage alcohol also vary significantly between age groups.

Although the frequency of *samogon* consumption rose with age, vodka consumption decreased. Members of the senior age group were half as numerous as the 30–49 years age group. As the result, the proportion of *samogon* in total alcohol consumption increased with age. Thus, in respondents aged under 30 it exceeded the proportion of vodka by 2.6 times, in the 30–49 years age group by 3.2 times, and among individuals over 50 years by as much as 9 times (as the standard of living decreases with age, older people have to consume less expensive alcohol beverages).

Fourth, respondents' indications of their motivation for drinking alcohol showed that consumption is now both more frequent and less orderly. Although the percentages differed according to region, the majority of respondents related their alcohol consumption to such traditional reasons as holidays or important family events (74–79%), to meeting relatives and friends (50–75%), and also to days off (29–37%). Other common reasons for drinking were ca-

TABLE 3.1. Frequency of Alcohol Use by Age (Percentage in Each Age Group)

Age (years)	Not more than once a month	Two to four times a month	Two or three times a week	Four times a week or more	Total
Under 20	—	33	67	—	100
20–29	—	28	44	28	100
30–49	—	25	15	60	100
50 and over	3.1	14.2	17.6	65.1	100
All ages	2	21	27	50	100

TABLE 3.2. Mean Alcohol Beverage Consumption by Survey Participants During 2 Weeks (Grams per Person)

Region	Total alcohol	Samogon	Vodka
Voronezh	1911	1586	325
Nizhegorod	1256	1032	224
Omsk	1768	1472	296
Mean	1644	1363	281

sual, occasional events like meeting acquaintances. Moreover, between 24 and 39% of respondents reported drinking without any particular reason, just because they wanted to drink. This indicates that for many respondents—including 45% of the men surveyed—consumption of alcohol beverages had become a habitual behavior, an integral attribute of everyday life, or a consequence of an already existing dependency on alcohol.

Fifth, frequency of alcohol use by itself describes neither the intensity of the spread of alcohol consumption among the population, nor its dangerous consequences for drinkers and for society as a whole. In most European and North American drinking cultures, regular use of low-concentration alcohol beverages generally does not result in widespread alcohol abuse and alcoholism. However, when the majority of beverages consumed in large amounts are undiluted, as in Russia, alcohol consumption may lead to serious negative consequences. According to data from this survey, rural Russians drink alcohol not only frequently, but also in large quantities (Table 3.2).

Table 3.2 shows that during the 8 days of each month when observation took place, survey participants consumed an average of over 1.6 liters of 40° alcohol, including 1363 grams of *samogon* and 281 grams of vodka—that is, 4.8 times as much *samogon* as vodka.

According to the study data, the individuals surveyed consumed an average of 3.6 liters during the 4 weeks of observation, including 3 liters of *samogon* and 0.62 liters of vodka. This amounts to an annual rate of 43.4 liters of alcohol beverages, including 36.0 liters of *samogon* and 7.4 liters of vodka.

Thus, high levels of alcohol consumption are typical of the rural inhabitants we studied. In absolute alcohol terms, each participating rural inhabitant was consuming 17.3 liters during a year, including 14.4 liters of *samogon* and 2.9 liters of vodka. Taking into account all age groups, including children and adolescents under 17 years, mean yearly consumption was 16.0 liters of absolute alcohol (13.2 liters of *samogon* and 2.7 liters of vodka).[1] As shown in Table 3.2, the highest level of drinking was found in the Voronezh region, where consumption was 20.1 liters of absolute alcohol per capita (compared to 18.6 liters in the Omsk region and 15.0 liters in the Nizhegorod region). In

[1]Among the overall population of Russia in 2001, per capita consumption was 13 liters of absolute alcohol.

Voronezh region, where sugar beet is grown, *samogon* production and consumption have a long and extensive history, and nowadays *samogon* accounts for almost 83% of alcohol consumption.

The Nizhegorod region's relatively low *samogon* consumption results from the fact that the drink has only recently become common, spreading during the antialcohol campaign of 1985–1988. The reduced availability of beverage alcohol during that period brought substantial *samogon* production to many parts of Russia where it had previously been produced only in small amounts or not at all. The observations made in these regions therefore support the view that existing traditions of *samogon* production and consumption are now a significant contributor to its widespread production.

It is not clear that all groups in the rural populations consumed alcohol in the same intensive manner. More detailed analysis of the different groups' attitudes toward alcohol showed that levels of alcohol consumption, especially *samogon* consumption, varied significantly depending on the gender and social status of respondents. For example, *samogon* consumption exceeded vodka consumption among the retired (by 15.4 times for men and 6.1 times for women); among industrial workers (by 6.8 times for men and 2.9 times for women); among male white-collar workers (by 1.5 times, whereas among females vodka consumption was higher); and among the unemployed (by 9.1 times for men and 3.6 times for women).

Sixth, these results are the outcome of uniquely Russian features in various aspects of drinking culture. The type of company, the reasons for drinking, the place and time of alcohol consumption, and the way of life and work of Russian villagers have produced a unique pattern of alcohol consumption.

To give one important example, public catering and drinking premises (cafés, beer bars, etc.) are undeveloped or absent in most rural areas of Russia. Consumption of alcohol beverages therefore usually took place at home (62–71% of respondents) or at acquaintances' homes (45–60%). For 13–16% of individuals, drinking took place in the street or in a public place (a term usually interpreted by rural inhabitants quite broadly, including drinking at work). According to the study's findings, older individuals preferred to drink alcohol in their own home while the young favored drinking at acquaintances' homes. For example, among the retired, 70–80% of drinking occasions take place in home settings. On the other hand, individuals under 30 years of age more often drank alcohol in the street or in a public place. Drinking alcohol in public places was almost 2.5 times less frequent among women than among men.

About two-thirds of respondents usually drank together with friends and acquaintances. Among rural inhabitants, drinking alone was a rather rare phenomenon (between 5 and 13% of all respondents). Among the retired the proportion was 14% and among the employed 8%. Females tended not to drink alone at all (2%).

The time of day when rural Russians drank alcohol presented a more unified picture. The majority of respondents (about 64%) drank alcohol during evening hours. At the same time, however, a significant proportion (37%) drank during the day. More than half of retired people started to drink at lunchtime. A majority of youngsters (54%) also drank alcohol at this time of the day—a subject of concern for the employed, 41% of whom fell into this category.

Seventh, in examining any specific drinking pattern, perhaps the most important consideration is the problems that pattern creates. In the study, more than half of the respondents avoided questions related to drinking problems, for fear of compromising themselves. Such a lack of openness is more typical of people who abuse alcohol than of those who drink it appropriately. Such people are usually more likely to overestimate their capacity to tolerate large quantities of alcohol. For that reason, many of the respondents when filling out the questionnaire refused to acknowledge being heavily drunk and having problems related to excessive consumption. Those who answered the questions cited the following problems arising from alcohol consumption:

- Severe hangover after drinking (nearly 66%).
- Increase in family problems (45%).
- Memory loss, losing balance when walking, and traumas related to falls (27%).
- Quarrels or fights with drinking partners (15%).
- Problems at work (up to 10%).

Given the delicate nature of the problem and the reluctance of many respondents to provide information, results in this area were far from complete. However, the survey organizers supplemented the responses with expert estimates and statistical data from all three regions, a technique which provided some interesting facts. According to crime statistics and Ministry of Internal Affairs estimates, 27–30% of all crimes are committed under the influence of alcohol. For certain crimes, primarily violent offenses, this figure is much higher (60–65%). Regions with high levels of *samogon* production see even higher levels of these "drunken crimes" in everyday life (70–75%).

CONCLUSIONS

Responses to the survey helped to illuminate several trends in Russian drinking culture, especially regarding *samogon* production and consumption.

First, the study showed there have been significant changes in the rural Russian drinking culture over the last two decades. Alcohol consumption has increased in frequency and volume; it has become both less controlled and more routine, which usually leads to high incidences of mass drunkenness and alcohol-related diseases; and it has increased among almost all the population subgroups, especially among females.

Second, state alcohol policy and the mass impoverishment of rural populations resulting from the uncontrolled reforms of the 1990s were shown to be primary contributors to the deteriorating situation. Economic and legislative measures have not contributed to stabilizing the alcohol situation, and indeed often hamper efforts in that direction. If in earlier stages of Russia's historical development the state's alcohol regulations helped to form a coarse culture of alcohol consumption, the current unreasoned policy seems to ignore interests of physical health and safety. It has provoked large-scale consumption of unofficial alcohol along with an unprecedented proliferation of counterfeit vodka that has increased lethal alcohol poisonings several times over. All this has further encouraged the widespread home production of alcohol beverages as a safer alternative to official beverages.

Third, the state has effectively lost control over the quality of alcohol beverages, an even more unfavorable situation resulting from its price and taxation policies. The high retail prices of officially available alcohol beverages, the rapid fall in rural incomes, and the abolition of legislative sanctions on the production and sale of *samogon* have all led to the consumption of cheap beverages made in home settings.

Fourth, *samogon* production has been exacerbated by the economic situation of the elderly. Many older people produce *samogon* as an ersatz currency that enables them to pay for firewood, home repairs, plowing, and even basic foodstuffs.

The almost complete suspension of state control over *samogon* production and of preventive and educational work among the population during the past decade has helped to increase *samogon* production and consumption. It has also led to passivity and reluctance to assist law enforcement efforts to end home production and sales of alcohol beverages.

The rise in the prevalence of *samogon* consumption in rural areas poses a real danger to Russian society by raising levels of alcohol abuse, weakening public health, and sapping the work ethic of the rural population. Such a situation requires a broad, scientifically grounded state program to curb drinking and alcohol dependency in rural areas.

Chapter 4

Local Alcohol Issues in Zambia

Alan Haworth

Although commercially produced beverages have been marketed for many years in most African countries, there is still a tradition of drinking homemade brewed and distilled beverages. None of this consumption is officially recorded. Many of these beverages are considered to be illicit, and hence their production and consumption are often concealed from the authorities. In many countries little is known of their importance to such issues as lost revenue, their impact upon social life, and the possibility of public health problems arising from their consumption. Because an extensive survey of drinking was carried out in Zambia 25 years ago, it was thought appropriate to carry out a further study, focusing this time on the consumption of these beverages, within the context of the drinking of alcohol generally.

It is being increasingly recognized that the consequences of drinking, at both the individual and societal level, depend not only on such simple parameters as quantity of alcohol consumed and frequency of consumption, but also on many other factors that comprise the pattern of drinking. From a public health perspective, such patterns involve both biomedical and psychosocial components. This chapter concentrates mainly on the latter, although reference is made also to some biological aspects. It was decided to examine, for example, the number and characteristics of heavy drinking occasions, the settings where alcohol is consumed, the activities associated with drinking, and the clusters of norms and behaviors directly related to drinking.

Zambia, a land-locked country in southeastern Africa, has a population of about 10 million people. A former British colony, it gained its independence in 1964. Unlike many African countries, which have largely rural populations, almost half of Zambians live in urban communities. The country is relatively large (some 752,600 sq km), and the population density in rural areas is generally less than 10 persons per square kilometer. Zambia's main source of income once was the production and exportation of copper, but because primary products have lost their relative value Zambia is now classed as one of the poorest nations in the world, with extensive poverty in both rural and urban areas. In recent years, some rural areas have been affected by acute droughts that have led to serious food shortages and to a lack of grain for the production of homemade beverages. Since the mid-1980s the whole country has also been seriously affected by the HIV epidemic: Approximately 23% of sexually active adults living in urban areas and about 14% of those living in rural areas are infected, but there is much variation from place to place. The expectation of life at birth has fallen from about 52 years to less than 37 years. Because Zambia is a land-locked country situated far from major markets, it has not been able to develop a highly industrialized economy, and unemployment is therefore widespread. There is instead a large informal sector, meaning that the actual income of many households goes largely unrecorded. The study to be described concerns the drinking of people living in three areas of Lusaka, the capital city. The question of extrapolating these findings to other parts of the city or to rural areas is discussed later.

The first question needing to be answered, however, is what is "unrecorded consumption," and how does this concept relate to that of being an illicit beverage? Many types of beverages consumed in African countries are described as "illicit," but whether this is a correct description will depend on the legislation regarding alcohol beverages in any particular country. In Zambia, all home-distilled beverages are illicit, but some home-brewed beer, made from traditional grains and produced in rural areas for home consumption, is not classed as an illicit beverage. The vendor of this same beverage in a designated urban area would, however, be liable to prosecution. Although definitions may appear to be clear to the authorities, they are by no means so to the producers and consumers of beverages. Even in rural areas, the fact that a homemade brew is being sold outside the home is often ignored; it is hardly worthwhile to attempt prosecution in such cases. In 1978 data were collected in Lusaka on the drinking patterns of a sample of its population in a collaborative study of Community Response to Alcohol-Related Problems (CRP) carried out simultaneously in three countries—Mexico, Scotland, and Zambia—and coordinated by the World Health Organization (Haworth, Mwanalushi, & Todd, 1981). Because all homemade beverages consumed in cities such as Lusaka are by definition also "illicit," this term is used for all such beverages in the following sections.

ILLICIT BEVERAGES

At the time of the CRP project, the list of recognized illicit beverages was relatively short but members of the public were not aware of the niceties of definition. The popular conception of an illicit beverage was (and remains) one that is "strong," either because of its high alcohol content or because of added ingredients that are usually thought to be dangerous. The main illicit beverages are either produced by fermentation or distilled. Traditional fermented beverages are usually described as "opaque" because they are unfiltered; they contain variable quantities of alcohol. The beverage is often drunk when the ethanol concentration is about 4%, although this will increase as time goes on because fermentation is allowed to continue. Yeast activity is terminated either because of the high concentration of alcohol, when fermentable sugars are depleted, or when spoilage by bacteria (i.e., acetobacteria) starts to dominate. Some commercially produced opaque beverages leave the brewery with a relatively high sugar content and an ethanol concentration of about 4%. Both commercial and illicit beverages are often left to "mature" at the outlet (sales point) until the concentration reaches about 6%. Depending on hygienic conditions in the place where the beverage is brewed and at points of consumption, the shelf life ranges from 2 to 5 days, when the concentration of volatile acids such as acetic acid makes the beverage unpalatable (at 0.03% m/m). Although the amount of alcohol and available sugars in factory beverages is regulated and should be strictly controlled, this is not the case for illicit beverages of any kind.

Types of Illicit Beverages

Kachasu is an illicit distilled spirit (one nickname for it is "pipeline") made in a two-stage process. The ingredients going into the initial brew include sorghum and maize (most commonly), sour (spoiled) beer, sugar, and yeast. The final product is sometimes graded, and the initial distillate (designated Number 1) contains the highest percentage of ethanol (but is also more likely to contain methanol). As the amount of water increases in subsequent distillations, the weaker Number 2 is produced. Number 1 is often mixed with a subsequent distillate. A product that is below usual quality or contaminated may be redistilled. The amount of ethanol is usually in the range of 20–30% by volume, although it is commonly believed that *kachasu* is the strongest beverage available, even by comparison with imported commercial spirits. *Kachasu* was and is still the main distilled illicit spirit sold in Lusaka, but its main distinguishing feature is usually exaggerated—as are its dangers—by nonconsumers. *Kachasu* (and perhaps other beverages) is sometimes called "cup-level," referring to whether a drinker thinks himself capable of drinking a full cup. It is also sometimes called *nsimbi,* which means iron. It is believed that other ingredients are added during "brewing"—including fertilizers and petrol. Most people do not understand or differentiate between beverages produced by the two processes: brewing and distilling. For example, the author of a report on the use and abuse of *kachasu* issued by the Zambian Drug Enforcement Commission (Mukuka, 2000) refers consistently to "*kachasu* brewing."

Before we began the present project, our initial impression was that older types of illicit beverages had been abandoned, and that one could rely on a rather simple classification in recording which beverages had been consumed. The situation turned out to be far more complex. Old-fashioned illicit beverages tend to be mentioned by older informants, but the places where they are said to be found are the longer established illegal settlements (some used to be called shanty towns). The names of these beverages are often unknown to young people, and when they have heard of them, their knowledge tends to be inaccurate. The beverages include types usually made from stale bread, sugar, yeast, and water—and, it is alleged, with occasional added ingredients to give them a "kick." The best known include *mbamba* and *skokiaan*. Another, *mbote*, is also called honey wine. Their usual alcohol content at the time of the CRP was 7–8%. But they remain on the market. Many have been dignified by the name "local wine," or they may just be known as "wine" as far as the purchaser, obtaining some from a private house, is concerned. They are often given a nickname (with very localized currency), because they are unlabeled. Sometimes the name may come from a prominent landmark (e.g., "tank" and "blue house") or from the name of the vendor (e.g., *Shims,* an abbreviation of a Zambian form of address, "father of . . ."), and sometimes the name is simply taken from the fruit or other substance used in flavoring, such as guava or tea. The mode of manufacture is essentially the same: A mix of the flavoring with sugar, stale bread, and yeast is fermented over a period of up to 72 hours. The resulting brew is then sold with an alcohol content of 7–8%, often on the next day, usually in measures of about 400 ml (in a plastic cup).

It seems that a small number of these beverages are prepared more as true wines, with a consequent increase in price. However, some of the prices mentioned indicate that there has been no essential change in the beverages except in public perceptions. Some of the so-called "local" wines may now have an alcohol content of up to 9%, but there is no clear comprehension of which are legally produced and sold and which remain illegal. Some years ago a home-produced "wine" called *akiki* was sold at shebeens (illegal outlets in private homes) and recognized as illicit. *Akiki* wine was said to have been "improved" (inquiries have not revealed how), and it is now apparently being sold more widely. At the time of writing, there is no evidence that it is obtainable from legal outlets, although this has been asserted. The lack of accurate information illustrates the current confusion as to the status of such beverages.

COMMERCIALLY PRODUCED BEVERAGES

In order to understand the place of illicit beverages in the drinking patterns of Zambia, it is necessary to know something about commercially produced and sold legal beverages. The most important is called Chibuku. This is a brand name given to the beverage thought most likely to be popular with the Zambian copper miners in preindependence days. It was deliberately manufactured to resemble the traditional home brew: unfiltered and "opaque" (a generic name given to this type of beverage). Many small breweries have now been set up that produce opaque beers, and they are often looked upon as just other forms of Chibuku. They carry names such as Chinika, Golden Brew, Makeni, Makungu, and Chat (a recent one), and they are produced and sold under official license. The producer of Chibuku (National Breweries, a subsidiary of South African Breweries) has attempted to regulate the quantity of alcohol available in its product by enforcing strict quality control. Because fermentation continues in the unfiltered product, it is usually released for sale when the alcohol content is less than 4% by volume and is not sold at above 6%. However, it has been suspected that other commercial opaque beers may have higher percentages of alcohol (up to 9–10%), and lax quality control may also allow beverages containing toxic products such as methanol to be produced. There is no evidence to confirm this allegation. National Breweries has proposed to the government that there should be a standard definition of opaque beers. Provided the regulations were properly enforced, this would act in the interests of drinkers as well as protecting the image and quality of the beverage. Because Chibuku is distributed in bulk carriers and then dispensed in plastic mugs holding 0.5 to 3 liters of beer, with drinkers sometimes sharing, assessing the amount drunk by any individual can be difficult. Some of the other opaque beverages, however, are packaged differently, such as in Tetra pack cartons, and hence purchasers may be more likely to drink the amount they have actually bought. It has long been a custom in Zambia to drink beer directly from the bottle. This is almost invariably more hygienic than using glasses, which are reused at outlets where the supply of any water (let alone hot water) may often be insufficient. It should be noted that in ordinary parlance, the word "beer" and its vernacular equivalents are commonly used for any beverage alcohol—even wines and spirits, and including illicit as well as legal beverages.

PREVIOUS STUDIES

There is virtually no information on the consumption of illicit beverages from previous studies in Zambia. A useful historical account may be found in Ambler (1992), and Colson and Scudder (1988) wrote a very valuable account of the drinking of the Tonga people living close to the Zambezi river in southern Zambia. Before Zambia gained its independence from the British, there was much hostility to restrictions on drinking in urban areas. Hence studies would have been difficult to carry out. Studies in rural areas focused more on production and sale than on individual consumption—for example, a study of drinking in Luapula Province (bordering on the Democratic Republic of the Congo) by Kay (1960). Some data were collected in Lusaka in 1969 as part of an unpublished study of psychiatric morbidity in primary health care, and it was noted that, combining both distilled and brewed beverages, illicit beverages comprised 6% of all alcohol beverages consumed by respondents in the two areas studied. This and other surveys carried out in the past 30 years have consistently shown that approximately two-thirds of men and one-third of women are drinkers, so that about 3% of the total population aged 15 and over would have consumed these beverages. The first systematic attempt to study the consumption was made in the CRP, but it was unsuccessful. At the time the fieldwork was being carried out, the local police scheduled a major operation to look for stolen goods, illicit distilling, and other illegal activities in house-to-house searches.

The most recent study overlapped with the survey reported here. The Zambian Drug Enforcement Commission baseline study of the extent to which the illicit alcohol *kachasu* is used and abused in certain urban communities attempted to assess whether there had been an increase in consumption (Mukuka, 2000). Not surprisingly, many of the 301 distillers and vendors of *kachasu* who were interviewed were somewhat reluctant to divulge information; 74.4% of informants stated that they knew that their operations were illegal. As Mukuka (2000, p. v) reported, "Although it was difficult to estimate the amount of *kachasu* brewed or distilled privately and sold . . . the study found that there had been a considerable rise . . . in the previous 2 or 3 years."

To a question on change, 41% of informants stated that consumption was increasing, 30% that it was decreasing, 20% that it was stable, and 11% stated that they did not know. However, a project based on focus-group discussions cannot provide reliable numerical data. Because the Commission's survey was contemporaneous with our survey, other results will be quoted at appropriate points in our account.

It has also been possible to tap one other source of current information: data collected in a large-scale study of HIV infection in couples, the Zambia/University of Alabama at Birmingham HIV Research Project (ZUHRP) (Allen et al., 2003; Haworth & Allen, 2001). Although the couples were interviewed before accepting to be tested for HIV, they could not be described as being absolutely typical of Zambians because an unknown proportion may have modified their drinking because of their HIV status. Basic information was obtained on their drinking habits at the beginning of the survey. Because it was collected by trained counselors, this information is likely to be as accurate as one might obtain in a situation not involving strict anonymity. Of the total of

4,642 participants, 95 (2.046%) reported drinking *kachasu* and 45 (0.969%) reported drinking other illicit beverages. When we carried out the CRP survey, considerable doubt was expressed as to how representative the sample was, in spite of using a carefully worked out sampling frame. In our report we specifically made reference to differences between communities (see Haworth et al., 1981). ZUHRP drew its sample from a different area of Lusaka, and this may account for the fact that none of these informants mentioned local wines. But the reason could just as easily be that the interviewers did not include mention of these beverages in their list of examples.

METHODOLOGY

Selection of Subjects

A general description of the conduct of a prospective diary study has been presented in chapter 2. This section therefore deals with aspects that were peculiar to the Zambian study. In light of our prior knowledge of likely current patterns of drinking, obtained in focus-group discussions, it was decided that our best strategy would be to look for a cohort of heavy drinkers. On the basis of previous experience this was not expected to be difficult to find. The following quotation is taken from the report of the Community Response Project (Haworth & Serpell, 1981), because it provides comparative data for the results of our survey and further justification for our choice of drinker:

> The picture . . . is . . . one of polarization . . . more than half claim to be total abstainers [and there are] regular, heavy drinkers who make up near half of the total drinking minority. This group of regular, heavy drinkers, which accounts for about 15 percent of the total population . . . is probably responsible for more than half of the . . . total consumption of alcohol. They drink enough to make them drunk at least two or three times a week. Two out of three of them are men over the age of 30, most of them married. The heaviest drinkers among them seem to be mainly over 50. We . . . were unable to detect any significant bias among them in educational attainment, type of job, or occupational status relative to the community as a whole. The close correlation between responses about drinking and those about drunkenness suggests to us that the norm amongst most Zambian drinkers is that one goes drinking in order to get drunk. Drinking . . . is largely a sexually segregated activity and scarcely ever involves a married couple drinking together . . . the men drink about 80 percent of the time in bars or taverns. Most of the heaviest drinking seems to occur in taverns, where large quantities of opaque beer are consumed gradually over periods of several hours. The small proportion of our sample who belong to the conspicuous minority, educated, urban sector of the population generally preferred to drink bottled beer in bars. This group seems to include a similar proportion of heavy drinkers to the rest of our sample. . . . Amongst the urban male drinkers an equally common occasion

> for the heaviest drinking of the month was termed a "drinking spree." These sprees feature as part of an urban male folklore in Zambia, and they often coincide with the arrival of a worker's monthly pay. Although a man's wife is never part of a drinking spree, other women may be. We found no evidence that the presence of women in a group of male drinkers has a moderating influence on the amount of alcohol consumed. . . . Drinking by (younger) women is also generally frowned upon. (pp. 10–12)

As a main objective was to study drinking patterns of individuals volunteering within households, it was decided to select only households where regular drinking was known to take place. Households were chosen in three locations. Two of them (Kaunda Square and Mutendere) were first established as "site and service" suburbs. When the government decided to prevent the mushrooming of "shanty towns" caused by rural–urban drift, serviced sites were established on which standard "extendable" houses could be built and individuals were allocated plots. There is thus great variability: Not all of the houses have electricity, the occupants have a wide range of occupations, and most houses are overcrowded. The third location, called Kamanga, was classified as "unauthorized," and the houses and local shops are generally ramshackle. It includes an area where illicit distilleries operate and where "skid-row" drunks can often be seen lying around on the ground at almost any time. Although this third location sprang up as a typical "shanty" compound, over the years the difference between the two types of area has diminished as Kamanga acquired services and better houses. Of the 75 households selected, 35 were in Kaunda Square, 25 in Mutendere, and 15 in Kamanga. Of 65 households where we were able to obtain details, only 45% used an electric cooker, with the remainder using charcoal. Even fewer had a refrigerator (37%), electric heater (9%—Zambia has very cool weather in winter), bicycle (10%), or videocassette recorder (15%), but 89% had a radio and 75% television. No questions were asked about family or personal income because this was felt to be too intrusive. As discussed later, many households must have had incomes higher than the income coming from wage employment.

Households were selected because of their accessibility to the research assistants supervising the entry of information in the diaries. Rather than focusing on the "skid row" population as such, we used a common means of identifying households in all three areas. Fifteen research assistants were enrolled, all of them working in various jobs at a health institution that includes a psychiatric hospital, a large paramedical training college, a primary health care center, a malaria research laboratory, and public health offices. In addition, one full-time research assistant looked after logistical and other aspects of the project, and one other supervisor was recruited for each area; these were professional health workers who had already had experience in other research projects. Each of the 15 field-workers was asked to include his or her own household as one of five in which diaries would be kept. After information was systematically obtained on the structure of each household, any drinking member was invited to keep a diary, with the assistance (supervision) of the research assistant where necessary, in return for which a payment would be made when the diary was handed in on completion of the project.

One hundred and twenty diaries were kept in the 75 households, by 88 men (73.3%) and 32 women (26.7%). Altogether, the households contained a total of 374 persons aged 15 or over (considered adults), or an average of 5 per household, a situation that is not uncommon in Lusaka. There were 69 male heads of household (61 with their wife) and 6 female heads, none currently with a spouse. These figures are a reflection of the devastation being caused by HIV/AIDS. In addition to the household heads and their children or nephews and nieces, there were 9 parents or grandparents, 56 siblings of the heads of household, and 29 others (friends or lodgers staying with the family).

Agreeing to keep a diary was of course entirely voluntary, and many members of households who were occasional or even regular drinkers might have declined. Of the 120 diaries, 56 were kept by respondents from Kaunde Sqaure, 21 from Kamanga, and 43 from Mutendere. All the female heads of household filled in diaries and 65 of the 69 male heads, along with 17 (22%) of the eligible offspring, 26% of the male relatives, and 16% of others, although none of the parents or grandparents did so. These proportions cannot be taken to represent the proportions of actual drinkers in the households. Details of their education and work status and religion and their status within the household are given in Table 4.1. Under religion, Christian churches were classified

TABLE 4.1. Demographic Data on 120 Diary Keepers

Occupational status	Number	%
Student	3	2.5
Housewife	5	4.2
Laborer	9	7.5
Artisan	34	28.3
Lower professional	20	16.7
Upper professional	2	1.7
Unemployed	16	13.3
Other	8	6.7
Totals	120	95.9

Education	Number	%
Primary	23	19.1
Junior secondary	27	22.5
Senior secondary	54	45.0
Missing—no response	16	13.3
Totals	120	100

Religion	Number	%
Catholic	58	48.3
Liberal Protestant	32	26.7
Strict Protestant	13	10.8
None	4	3.3
Other	7	5.8
Missing	6	5.0
Totals	120	99.9

Status in household	Number	%
Head	71	53.3
Spouse	12	15.0
Children	24	20.0
Other relatives	14	11.7
Totals	121	100

as Roman Catholic, strict Protestant, and liberal Protestant. The strict Protestants tend to forbid any sort of drinking among their members.

The ages of diary keepers ranged from 19 to 60 years among males (mean 33.7) and 19 to 52 years among females (mean 33.8). Assessments of the number of young regular drinkers in Zambia have been attempted only in school and student populations, and none can be said to have given a picture of the general population. With a high dropout rate from school, many people in this age group are not easily accessible for surveys. In the CRP survey, we likewise found very few respondents under the age of 20.

Data were collected during six 5-day periods selected to include Christmas and the New Year as well as weekend, month-end, and midweek periods. Overall, data were collected on 30 days between 24 December 1999 and 29 February 2000. The diary consisted of a small school exercise book with a fold-out page pasted inside the front cover giving the following instructions:

> **Information to be put in the Diary**
>
> Date and Day about which the entry is being made. Note: For our purposes, each day starts at 06.00 hours. Any drinking before this time will belong to the previous day, and will not be counted as part of the present day. If the drinking took place in more than one place, record each occasion separately, with the times and all other details.
>
> 1. About what time did drinking start, and about what time did it end.
> 2. Where did the drinking take place. Give *exact* details.
> 3. Who were you drinking with most of the time (e.g., alone, with friend, brother-in-law, etc.).
> 4. What did you drink (give exact name or type of drink, e.g., *Mosi*, *Chibuku*, Whisky Black or *kachasu,* etc.)
> 5. How much of *each type* did *you yourself* drink.
> 6. About how much did *you* spend altogether.
> 7. What effect did the drinking have (e.g., got drunk, just all right, etc.).
> 8. Was this a special occasion, and if so, what.
> 9. Did the drinking lead to you having any problems; please describe.

Where a drinker was not literate in English, the instructions were translated ad hoc by the research assistant, who reviewed the entries very carefully initially in order to ensure accuracy. Most of the respondents, however, were competent in English; 101 (84%) also used English in answering two additional questionnaires not discussed here.

Classification of Beverages

It was anticipated that problems would occur with regard to Chibuku, the main commercially produced opaque beverage. Aware that it is continuing to ferment, customers insist on not drinking new deliveries. We found by collecting a total of 12 specimens that the average alcohol content at time of sale was 5.7%, and we used this figure in calculating the number of units consumed. However, this was not necessarily always the case. Chibuku is usually sold in either half-liter or 1-liter units, and we also checked that customers were not being sold short. Short selling rarely took place as far as we could tell. Note that the instructions for keeping the diary refer very explicitly to the amount taken by the drinker himself or herself, and research assistants were instructed to check upon this. We were informed by their supervisors (one supervisor to five assistants) that this was regularly done. However, containers of Chibuku are often passed around a group of drinkers, and it proved impossible to be sure of the exact amount consumed by individuals. From subsequent focus-group discussions and data gathered subsequently on the drinking of Chibuku, it is apparent that the total amount purchased rather than individual intake may have been recorded on many occasions. We assumed that where a group regularly met to drink together, it was unlikely that any individual would be allowed to drink much more than he had actually purchased; this was confirmed in focus-group discussions. Assigning one unit to a bottle of lager, each liter of Chibuku was likely to have comprised almost four units but would be shared with one or more drinking companions. We therefore assigned 1.5 units to each liter of Chibuku reported as shared. We also had to make some other assumptions. All wines designated as local wines we classed as illicit whether the informant had done so or not; this detail was usually absent from the diary. A small amount of bottled commercial wine was also consumed by a few of the drinkers. Where other opaque beverages were consumed, a check was made on the alcohol content of a sample and this was used in calculating total consumption.

Terminology

Regarding the terminology used in our report, we decided on a number of conventions. All clear beers and ciders produced commercially and sold in bottled or canned form are listed as lagers; Chibuku produced by National Breweries is listed as such; other opaque beers are listed as "Other opaque"; *kachasu* includes any other illicit distilled beverage; all other illicit beverages, including various local "wines," are listed as "Other illicit"; and commercial wines and spirits (mostly imported, or blended in Zambia) are listed as "Eurowine" and "Eurospirit," respectively, following a convention used in previous studies in Zambia. We also included an "Other" category for the occasional beverage that does not fall into any of these categories, such as *amarula*, which is a winelike beverage made from a local fruit.

Our solution to the question of defining a drinking occasion was to ask if the drinker moved to drink somewhere else (and possibly with other companions). If so, a full set of questions was then to be answered according to the list. We allowed for up to three drinking occasions to be described in any 24-hour period. Of the total number of 2,882 drinking occasions reported by all the respondents during the 30-day period, 93.8% were single occasions, 5.4% were second occasions, and 0.76% were third occasions.

RESULTS

Preferences and Patterns of Consumption

In this section we begin by reviewing the consumption of different types of beverages in terms of frequency of drinking (see Table 4.2).

Lager and opaque beer were the most popular beverages, and illicit beverages were consumed by 35 respondents (29%). It has been remarked that opaque beer is so cheap that it is hardly necessary for drinkers to seek illicit beverages, although many have no objection to drinking them when the chance occurs. In the past there tended to be distinctions between drinkers from different income groups; for example, those with the lowest income drank illicit beverages, those in the next group drank Chibuku, and those with the highest income, depending on actual cash available, drank local or imported lagers, wines, and spirits. This is still observed to some degree, as in the differences in preferences recorded at the three study sites (Table 4.3). One of the major points made by Mukuka (2000) in his survey of *kachasu* distillers was that the manufacture and consumption of illicit beverages were a product of poverty.

As noted in earlier surveys, there was a shift in the type of beverage consumed during the course of a month, depending on the amount of ready cash available. Although this change is not as clear in this group of drinkers, some of those who drink illicit beverages may do so because of lack of ready cash.

Some beverage preferences are likely to go together. For example, of the 108 individuals who drank opaque beers of any kind, 77 drank Chibuku alone, 2 drank other types of opaque beer alone, and 29 drank both. However, there is

TABLE 4.2. Preferences of Drinkers for Various Beverages

	Males		Females		All	
Beverage	Number	%	Number	%	Number	%
Lager	77	87.5	29	90.6	106	88.3
Chibuku	81	92.0	25	78.1	106	88.3
Kachasu	21	23.9	3	9.4	24	20.0
Other illicit	14	15.9	4	12.5	18	15.0
Eurospirits	12	13.6	7	21.9	19	15.8
Eurowine	7	8.0	7	21.9	14	11.7
Other opaque	23	26.1	8	25.0	31	25.8
Other	1	1.1	5	15.6	6	5.0

TABLE 4.3. Beverages Consumed at the Three Study Locations

	Kuande Square		Kamanga		Mutender	
Beverage	Number	%	Number	%	Number	%
Lager	46	82.1	21	100	39	90.7
Chibuku	47	83.7	20	95.2	39	90.7
Other illicit	15	26.8	2	9.3	7	16.3
Eurospirits	12	21.4	—	—	7	16.3
Kachasu	11	19.4	5	23.8	8	18.6
Other opaque	12	21.4	5	23.8	14	32.6
Other	3	5.4	1	4.8	2	4.7
Eurowine	5	8.9	3	14.3	6	14.0

much less overlap with regard to illicit beverages. Of the 35 who drank any kind, only 4 mentioned drinking both *kachasu* and other illicit beverages. There is also little overlap between eurowines and eurospirits, with only 4 of 29 drinking both.

With regard to the consumption of *kachasu*, although many of the regular drinkers belong to an older age group, sellers of *kachasu* (Mukuka, 2000) indicated that a large proportion of their customers were from a younger age group. Male youths (not defined) were described by 47.6% of sellers as normally using or abusing illicit alcohol, versus 32% who stated that male adults were the normal users. The diary study confirms that a high proportion of the older age group (37–60 years) drank *kachasu*, compared with younger age groups, but the youngest group (19–27 years) outnumbered the middle group (28–36 years). On the other hand, the most numerous customers were not necessarily the most regular. In terms of average quantities consumed, the three age groups consumed nearly equal amounts of opaque beer (including Chibuku), whereas the middle age group not only consumed illicit beverages less frequently but also consumed smaller average quantities (8% of the illicit wine and 24.5% of the illicit spirit).

Quantities of Individual Beverages Consumed

Opaque beer was the most popular beverage, accounting for 42.6% of all alcohol, including 14.6% for "Other opaque." Lager accounted for 32.3%, illicit beverages accounted for 12.6%, and other beverages, including commercial wines and spirits, for only 2.35% of the total. The last figure is somewhat surprising in the sense that the data collection period included both Christmas and the New Year, when some diary keepers might have celebrated by drinking these more expensive beverages. But we have already remarked that the general pattern of drinking seemed to change hardly at all in this group.

Although the information provided by "usual" and "highest" drinking days might be useful in examining the potential for harm for lighter or moderate drinkers, the constancy of the intake in these drinkers is a main indicator.

Although the actual numbers were small, distinct patterns of beverage preference may be seen that are consistent with previous experience, although in somewhat attenuated form. Differences between men and women were marked. Women reported drinking lagers more than Chibuku but tended to drink less of both than men, although many were also heavy drinkers. There was a large difference between the average total amount of *kachasu* drunk by men compared with women. More commercial spirits were drunk by men and somewhat more wine by women, although the numbers were too small to draw definite conclusions.

Again, with a caveat regarding the small numbers, some not unexpected patterns emerged with regard to age. Both youth and the older age appeared to be associated with drinking larger amounts of *kachasu* and other illicit beverages, thus partially confirming the information obtained by Mukuka (2000). The total amounts of Chibuku and other opaque beers consumed were little affected by the drinkers' age. With regard to education and job status, the pattern was the same as that described earlier in the CRP survey—a tendency for more education and higher job status to be associated with "high-class" beverages.

It seems that illicit beverages were often drunk out of necessity rather than for preference, as they are more affordable to the young and the old, who may find more socially acceptable beverages just out of reach. In addition, the young (and especially adolescents, who may have been unwilling to keep a diary under the watchful eyes of parents or guardians) may well have sought beverages obtainable at the less public outlets. This was subsequently confirmed in a focus-group discussion with a group of male adolescents. Merely recording that a particular beverage has been consumed over a specific period cannot indicate its importance in the drinking program of an individual. In the analysis of such data it is often useful to construct a frequency–quantity index (which could be done in any further work on the data obtained in the present survey), but some patterns are clearly apparent without this elaborate processing. Although only a minority of drinkers drank *kachasu*, the quantity consumed by the male drinkers was not insignificant and, besides being far greater than the average amount consumed by females, also exceeded the male drinkers' average lager consumption.

Total Quantities Consumed

In this and the following sections of this chapter the intention is to describe some aspects of the general context of drinking. Comments will be made on the relevance to illicit beverage consumption where possible. Because of the considerable problems found in calculating quantities and the many assumptions that have had to be made, we present the data with some hesitation. Since the completion of the survey, a technique has been developed that uses samples of the actual containers in which opaque beer is served and that helps the individual to calculate the amount he or she would personally have drank. However, a project now in progress has not yet provided sufficient data for any corrections (if necessary) to be made.

In the CRP study the quantities consumed were noted to be exceptionally high in comparison with other countries. A total of 306 respondents gave information about their last-day drinking, and 54% of 213 males and 40% of 93 females reported drinking 9 or more standard units. The average amount was 11.0 units (11.8 for males and 9.2 females), and the total during the previous week was 15.7 standard units.

The mean daily intake of alcohol beverages can thus be calculated to have been 9.3 units, based on all drinkers and all 30 days. The actual mean amount based on the diary entries and using only the actual drinking occasions as the base for each mean gives considerably higher mean daily intakes. When averages were taken over 5-day periods, there was little variation, and the daily amounts were about one-fifth more than those recorded in the CRP survey (see Table 4.4). Recall that the average concentration of alcohol currently found in opaque beers is higher than at the time of that study. In the present study, Christmas Day was also a Saturday, making the higher average level of consumption unsurprising; in fact, what was perhaps remarkable was the relative constancy of the amounts consumed. As might be expected, people tended to drink more on Saturdays, whereas midweek drinking was relatively less. In deciding on a definition of "weekend" drinking, the data from this small sample are somewhat ambiguous regarding Friday and Sunday drinking.

The Cost of Alcohol

We calculated the price of alcohol beverages in comparison with the cost of other commodities and also with the average incomes of the types of people

TABLE 4.4. Mean Consumption of Alcohol in Different 5-Day Periods

Period	Mean	Standard deviation
24–28 December 1999	57.092	38.442
31–04 January 2000	57.883	40.753
17–21 January 2000	47.333	38.794
25–29 January 2000	47.783	37.949
10–14 February 2000	56.050	42.418
25–29 February 2000	58.858	50.342

from whom we obtained information. All rates are quoted in U.S. dollars and cents. It is evident that some forms of beverage are exceptionally cheap, because there is only a small cost involved in bottling or packaging and the short shelf life means that there is a very rapid turnover of products. At the time of data collection, the cost of various commodities was as follows: average size loaf of bread (there are no standard sizes), 25–30 cents; cooking oil (2.5 liters), $3.00–3.50; sugar (2 kg), $1.25; maize meal (25 kg), $6.00. Maize meal is the dietary staple; 25 kg will feed a family of two adults and two children for 3 weeks. Most lagers cost about 50 cents per bottle (usually 375 ml), and opaque beers about 30 to 35 cents per liter. Imported beverages are much more costly; prices commonly ranged from $10 per bottle for imported wines to $20 or more for spirits. Some local commercial wines could be purchased much more cheaply. Working out the price of illicit beverages was difficult because of the variety of bottles and containers in which they were sold. Prices were usually 20 to 30% less than for commercial opaque beer, based on the amount of alcohol likely to be present. However, if a beverage was sold by the cup, the price was sometimes more than that of Chibuku, by a similar margin. *Kachasu* and similar spirits were usually cheapest, and a 375-ml bottle could cost as little as 20 cents. The average amount spent at Christmas and the New Year was somewhat higher than during other periods—$2.65 and $2.01, respectively. About 40% of respondents spent $1 or less each drinking day during the Christmas/New Year period and 43% about $1 or less during the other periods.

Drinkers are on the whole very canny as to whether they are getting value for money. According to our calculations, a small number of illicit beverages allow consumers to drink more alcohol per unit price than any commercial beverage. However, the consensus appears to be that drinking Chibuku is still the cheapest way to get drunk. At the time of the CRP, it was said that one could get drunk for less than the price of a loaf of bread. That is still the case. These prices must be set against the incomes of the purchasers. Most of the respondents were likely to have been earning less than the equivalent of $60 per month in declared income. We did not ask any questions on income because past experience had shown that we were unlikely to obtain accurate answers. Although information was available on a wide range of likely incomes from official sources, many individuals or households had additional sources of income as well. Those living in rented accommodation would pay about $10 per month, and although most of the diary keepers did not come into this category, it should be noted that there were 29 "lodgers" among the 120. As individuals they may well have had fewer calls on their income while at the same time augmenting the income of the households where they stayed. From the figures given, however, it appears that the majority of the diary keepers were spending amounts beyond their means in keeping up their drinking habits.

Drinking Places

It was remarked at the time of the CRP study that the laws governing drinking outlets were often disregarded. That situation has persisted and has had a major influence on styles of drinking within Lusaka. We can refer only to Lusaka in this regard because the licensing of drinking premises and the enforcement of the local regulations are in the hands of local authorities. The word "tavern" has always been somewhat ambiguous. At the time of independence in 1964, the old beer halls were closed and new municipal taverns opened. Although the beer halls were able to accommodate well over a thousand drinkers, the taverns could serve several hundreds in hardly more salubrious surroundings. A recent television documentary from Lusaka showed scenes of completely decrepit former municipal taverns that have been taken over by private entrepreneurs with the objective of making as much money, with as little outlay, as possible. They are now largely empty of customers. Soon after independence many "tea rooms" serving beer were opened, but it was evident that they would soon become either bottle stores selling mainly lager, or private taverns selling mainly Chibuku, although they increasingly sell lagers as well. The current legal framework for defining illicit beverages and for controlling the manufacture, distribution, and sale of legal beverages is complex and out of date. For example, the Liquor Licensing Act (Chapter 167 of the Laws of Zambia) makes reference to miles even though Zambia uses the metric system, and it uses the phrase "six o'clock in the evening" when Zambia officially uses the 24-hour clock. An intoxicating liquor is defined as having 3% proof spirit. In addition to the Liquor Licensing Act, the Traditional Beer Act remains in force. According to the first Licensing Act no person under the age of 18 years is allowed to drink (in specified circumstances), but under the Traditional Beer Act anyone aged 16 years or more may drink Chibuku. Because this beverage is now almost invariably sold with a higher concentration of alcohol, the whole purpose of age restriction is defeated. The regulations referring to "off-license" premises still prohibit drinking at or near to these outlets, but in fact the trend within Zambia has been to set up tables, provide shade from the sun, and even offer music and entertainment. The Markets Act forbids the sale of alcohol beverages within markets, and there is a specific provision for a license to be refused (in terms of Chapter 167) if it is considered that there is no established need for yet another outlet. Even so, outlets proliferate in and especially around markets.

In addition to the most used types of places listed in Table 4.5, there were 71 drinking occasions at hotels, 69 at night clubs, 12 at a party, 7 at work, 7 at a funeral, and a further 71 elsewhere. The numbers of respondents never using various places are of some interest. Of the common drinking outlets, shebeens are the least frequented; only 27 respondents reported using them at all, and of these, 11 drank there only once or twice. This is consistent with the number

TABLE 4.5. Types of Drinking Places and Frequency of Use During Past 30 Days

Type of drinking place	Never	1–4 days	5–11 days	12–28 days	Total days	Average/ day
Tavern	42	28	27	23	716	24.53
Bottle store	29	27	27	27	841	28.81
Bar	56	44	16	4	268	9.18
Home	35	40	33	12	732	20.55
Shebeen	93	11	1	1	125	4.28

taking illicit beverages. Likewise, bars too were not very popular and were visited only occasionally, if at all.

With so little enforcement of licensing laws, an expansion of shebeen outlets was hardly necessary, and in any case *kachasu* and wine could be sold for off-premises consumption. The word "tavern" is also ambiguous in another sense, because there is often little to distinguish a private tavern from a shebeen. Drinking in hotels and night clubs is generally the preserve of the better off, but at some hotels drinking takes place when people attend meetings utilizing hotel premises. Details were given earlier of the comparative cost of drinks served from different outlets. Because so few of our diary keepers drank at the more expensive outlets, we have given only the minimum prices charged for commercial beverages.

Regarding drinking outlets (see Table 4.6), 65.6% of women made no report of drinking at a tavern (vs. 23.9% of men), whereas 37.5% of the men reported never drinking at home versus only 6.3% of the women. Both men and women drank at least once at a bottle store in more nearly equal proportions (78.4% men and 68.8% women). It has to be kept in mind that beverage preference also influences where a person drinks; a detailed analysis of a sufficient sample would have to take this into account.

We have included some special "occasions" with the drinking places in Table 4.6. It is noteworthy that so few "partics" were mentioned for a period

TABLE 4.6. The Gender Divide in Use of Drinking Places (Percentages)

Type of drinking place	Males	Females	All respondents
Tavern	76.1	34.4	65.0
Bottle store	78.4	68.8	75.8
Bar	62.5	28.1	53.5
Home	62.5	93.8	70.8
Shebeen	28.4	6.3	22.5

including Christmas and the New Year. It is also surprising that no "kitchen parties" were mentioned by women. These are held for the bride-to-be before a wedding and are ostensibly for the purpose of allowing the older women to give advice. It is traditional for beer to be offered during funeral wakes; at least 70% of the respondents would expect to attend a funeral during the year.

Drinking Companions

The figures in Table 4.7 suggest that there has been little change in choice of drinking companions over the period of a quarter of a century since the CRP study. Drinking with a spouse or children is uncommon, even within the home. The most common style for men is to drink at a bottle store or tavern with friends and for women to drink at home with relatives or friends. Drinking at "kitchen parties" has only become popular since the time of the CRP; because we did not ask specifically about this type of event it must be assumed that this is where some of the women took larger quantities of alcohol. Note that the terms *relatives, in-laws,* and *family* refer respectively to parents, brothers, and sisters, those related by marriage, and the sons or daughters of the respondents.

As regards drinking with the spouse, although only 1.8% of men and 7.85% of women reported doing so on all occasions, 43.8% of women and 17% of men reported drinking at least once with their spouse during the 30 days of the study; in other words, some men do drink occasionally with their wives. There has been little change in this regard from the style of drinking almost a quarter of a century ago. We noted that drinking with the spouse was more common at home than elsewhere in the CRP survey but that it was nevertheless an infrequent activity. Recall that a number of the women drinkers (heads of household) were widows so that the proportion drinking with a spouse was actually slightly higher, taking this fact into account. No diary keeper

TABLE 4.7. Reported Companions on Drinking Occasions (Male, Female, and All Respondents)

	Females		Males		All respondents	
Drinking with whom	Occasions	%	Occasions	%	Occasions	%
Alone	105	13.51	301	14.07	406	13.92
Friends	825	54.7	1496	69.94	1921	65.88
Relatives and friends	46	5.92	92	4.30	138	4.73
Relatives	91	11.1	107	5.00	198	6.79
Spouses	61	7.85	39	1.82	100	3.43
In-laws	38	4.85	72	3.37	110	3.77
Family	5	0.64	12	0.56	17	0.58
Other	6	0.77	20	0.94	26	0.89

reported failing to drink with a friend at least once during the 30 days, and a quarter of the respondents had never drunk alone. Although drinking among the study group is a very social activity, there was a definite pattern as to whom people drank with.

Time Spent Drinking

The amount of time spent drinking can be influenced by a number of factors. Our very limited sampling of various days and other periods (weekend, end of month, etc.) can give only an impression, and it corresponds with what might be expected, although the differences were not very large. As noted earlier, there was some increase in the number of occasions associated with the festive season (which hence influenced the average time spent drinking). Although women tended to drink for shorter periods, a large proportion of the drinking women drank for long periods (5 or more hours) on Christmas Eve (Christmas is not necessarily a time for cooking elaborate meals). Taverns were the setting for long, heavy drinking sessions. As might be expected, higher proportions of respondents drank for longer periods on weekends, while midweek drinking occasions were briefer but still lengthy in comparison with lighter drinkers. Other possible influences on the amount of time spent drinking are illustrated in Table 4.8.

Table 4.8 shows that women drank for much shorter periods than men, which is probably why shorter drinking times were reported for drinking at home. The actual number reporting lengthy shebeen drinking was too small to

TABLE 4.8. Average Daily Drinking Time Over 30 Days, by Sex, Place, and Drinking Companion

	1–3 hours		4–5 hours		More than 5 hours	
Parameter	Respondents	%	Respondents	%	Respondents	%
Males	26	29.5	26	29.5	36	40.9
Females	15	46.9	12	37.5	5	15.6
Location						
Tavern	23	29.5	25	32.1	30	38.5
Bottle store	35	38.5	25	27.5	31	34.1
Bar	21	32.8	23	35.9	20	31.3
Home	34	40.0	29	34.1	22	35.9
Shebeen	7	25.9	6	22.2	14	51.9
Drinking with						
Alone	35	38.9	27	30.0	28	31.1
Friends	41	34.2	38	31.7	41	34.2
Friends/relatives	14	23.3	22	36.7	24	40.0
Other relatives	23	38.3	24	40.0	13	21.7
Spouse	10	34.5	15	51.7	4	13.8

draw any firm conclusion. It seems that, in general, companionable drinking with friends was preferred by the majority and that if drinking took place with others, it was of shorter duration.

Problems Arising From Drinking

The CRP survey had already demonstrated that many Zambians do not see heavy drinking as a problem in itself, and many do not directly link drinking to medical or social problems. Drunkenness was not in itself seen as problematic, and drinking was often not described as *real* drinking unless it involved some measure of experienced intoxication. Only 24% of the diary keepers did not describe themselves as getting drunk. In the 2,477 drinking occasions described, 1,574 involved getting drunk and 402 getting very drunk. This conforms to what has been described as the Zambian style of drinking: drinking with the intention of getting drunk. Binge drinking remains the pattern. About one-fifth of the diary keepers were definitely drunk on about three-quarters of the drinking days recorded, and about half on at least 2 days per week.

In examining the number of problems reported, it must be remembered that comparisons can only be made with global assessments of the occurrence of problems over much longer periods. For example, in response to a question on whether an individual has been injured in the last year or any time before that year, the apparently low rates of the occurrence of problems may be an artifact.

The data in Table 4.9 show that only a minority of respondents experienced self-defined problems linked to drinking during the 30 days recorded. Findings from the CRP using data collected in the accident department of a large hospital indicated that 14% of those with alcohol-related injuries were

TABLE 4.9. Spontaneously Reported Problems Occurring Over 30 Days: 2,477 Drinking Occasions (Numbers and Percentages of Total Drinking Occasions)

	Times experienced by individual respondent				
Reported problem	Once	Twice	Three times or more	Total occurrences (occasions)	Percentage of all occasions
Falling	23	3	—	29	0.98
Fighting	66	20	1	115	3.88
Injury	9	—	—	9	0.30
Feeling ill	57	11	1	86	2.90
Locked out	28	—	—	28	0.94
Hangover	48	19	7	109	3.68
Money cost/spent	12	3	—	15	0.51
Other	?	?	?	97	?

likely to be repeat attendees (Haworth, 1988). The three largest categories in the present study (fighting, feeling ill, and suffering a hangover) were also more likely to be experienced on more than one occasion, but in the case of suffering a hangover only seven reported this as occurring on three or more occasions during the 30 days. When the AUDIT questionnaire (Babor & Grant, 1989) was administered after the diary stage of the project, it was found that 30% reported a morning drink at least weekly, and 35.6% stated that they had a memory blackout at least once per week.

We did not systematically examine the relationship of drinking of any particular type of beverage to the occurrence of ill effects, although this should be done in any future survey. We did, however, specifically examine *kachasu* in relation to problems (of total alcohol consumed, *kachasu* contributed 7.27% of problems and other illicit beverages 2.31%, for a total of 9.58%), and we found that this bevereage seems to be less associated with admitted drunkenness than other illicit beverages (see Table 4.10). Although caution is needed in drawing conclusions because of the small numbers, the link between its consumption as seen in this study and either being older and tending to drink less, or being young and also tending to drink less (possibly because of *kachasu*'s reputation), may offer an explanation. It will be necessary to carry out a more focused study in the future.

DISCUSSION

As already stressed, the sample surveyed in the CRP project could not strictly be considered representative of Zambia as a whole. Yet from our observations of drinking elsewhere in Zambia, as well as from observations in routine clinical work, it is apparent that the survey gave a close approximation to describing the styles of drinking common in Zambia at the time and the problems consequent on some of them. It was certainly sufficient for policy formation.

TABLE 4.10. Problems Associated With Drinking Kachasu and Other Illicit Beverages on 2,477 Drinking Occasions

Reported problem	Total occurrences	With kachasu		With other illicit beverages	
		Number	%	Number	%
Drunk	1574	93	5.0	271	17.2
Very drunk	402	25	6.2	29	7.2
Fighting	115	8	7.0	18	15.6
Hangover	109	1	0.9	20	18.3
Illness	86	1	1.2	12	14.0
Injury	9	1	11.0	3	16.6

However, a major gap in knowledge of the consumption of illicit beverages needed to be filled. The current survey was formulated with that objective in mind, but it has to be asked whether the diary keepers represented any population other than themselves. We might call the sample a "modified snowball" sample: Research assistants known to come from families with regular (if not heavy) drinking were used to recruit families known to them, with similar drinking habits. Because the objective was to examine illicit drinking in the general context of drinking, and because it was known that this group was likely to yield a proportion of such beverage drinkers without being a distinct group of "skid-row" type drinkers, we decided on the method of recruitment described. It is certainly of interest that the broad picture of drinking styles has not changed substantially over a period of almost 25 years, and this tends to confirm the representativeness of the sample. Although the regular drinkers of *kachasu* frequenting shebeens might have been approached, they would have formed a much more highly selected group. If we had simply attempted a general population survey we might have found a very much smaller proportion of drinkers of illicit beverage, as is indicated by the other data we have quoted. The data from other surveys indicate that we found a much higher percentage of informants who could tell us about their consumption of illicit beverages than was likely in any other survey so far carried out.

Our post hoc focus-group discussion did not enlighten us sufficiently, but if such a discussion had preceded a preliminary survey carried out for the specific purpose of recording how Chibuku was drunk and paid for, as consumed in a group setting, and then had fed this information into specific instructions to research assistants on the checks they should make, it is likely that we would have gained a more accurate picture. We have already referred to a modification of the method. The amounts of alcohol recorded as being consumed, although large, are not dissimilar to those recorded in the CRP project and by heavier drinkers in other surveys. As discussed in chapter 2, the prospective diary method has been shown to yield higher alcohol consumption than other methods. But studying such a population, on which we have prior data from other studies, allows us to make comparisons and certainly does not invalidate differences found within the population. This relates specifically to beverage preferences, the location of drinking, drinking companions, and also the frequency of occurrence of problems related to drinking. One of the major debates regarding the unwanted effects of alcohol concerns the concept of alcohol-related problems.

The comparison with other studies clearly indicates that the current study population contained far heavier drinkers than the population at large. They also admitted to experiencing more problems, but it is impossible to quantify the increase in proportion to the quantities of alcohol consumed. Is the problem one that is defined by an external agency or investigator, or should the problem be defined in terms of what the drinkers themselves believe? Our data from the two studies, covering almost a quarter of a century, would seem to indicate that attitudes and perceptions of drunkenness have altered very little. Some aspects of drinking have not changed, and one in particular could be highly significant at the present time. Men tend to drink with their friends and not with their spouses. Because so much drinking takes place away from the home (although rarely, probably, very far away), there are likely to be opportunities for meeting sexual partners and thus promoting the spread of HIV infection. However, the relationship of drinking to the transmission of HIV is complex.

Public perceptions of some alcohol-related problems tend to be at variance with those of the investigator. It is assumed by many that drinking illicit beverages is more likely to lead to health problems. This small study has not confirmed that impression and specimens of some of the beverages being consumed did not contain harmful amounts of toxic substances (see chapter 11). This is in accord with earlier analyses for both chemical contaminants and aflatoxins, carried out in 1977 (Lovelace & Nyathi, 1977). There can be no doubt that many homemade beverages are produced in less than hygienic conditions, although the regular vendor of *kachasu* or local wine wishes to keep her customers, not kill them. It has been argued that *kachasu* especially should be collected from licensed distillers and then sold commercially and legally, say as "Zambia gin" (Reilly, Nwegbe, & Ofafir, 1974). We have drawn attention to the probable lack of enforcement of quality control regulations and the manifest lack of enforcement of regulations regarding drinking outlets in Zambia.

There also appears to be a lack of appreciation on the part of the public regarding which particular beverages are officially classed as illicit. This needs more exploration. A picture thus emerges of a system that is now much more anarchic than before and with potential for continued harm arising from beverage alcohol consumption. It should be noted that the government is probably also the loser in that taxable revenues consistent with the actual production and consumption of alcohol are not being obtained. It is often suggested that high taxes on alcohol serve the dual purpose of bringing in much needed revenue and also acting as a brake on excessive drinking. Cheap beverages are widely available, and the price of the cheapest legal beverage is not much more than the price of many illicit beverages. Although the licensing laws theoretically prevent the proliferation of legal outlets, there has been a marked increase, especially around markets. The young are less able to patronize such outlets. We showed in the CRP survey that older customers would wish to chase away the youth, not so much because they disapproved of their drinking as that they were embarrassed by their company. The youth need to seek outlets that sell beverages they can afford and in circumstances where there is less exposure to the public gaze. Members of the public when asked about problem drinkers in Zambia almost invariably mention "youth."

There is an increasing number of street kids—many of them AIDS orphans—on the streets of Lusaka, and many are seen drunk during the day. A recent study (Haworth & Chita, 2002) has drawn attention to alcohol being drunk especially by young girls engaging in prostitution.

Although "old-fashioned" illicit beverages such as *mbamba* are still available, there seems to have been a trend toward a change of name to "wine." However, *kachasu* remains the same in all essentials. Although it was noted that education tends to be associated with a wish to drink bottled beer, opaque beers still appear to have been very popular in the population studied. Home-brewed opaque beer sold in town would be illicit and other opaque beers sold from unlicensed premises would also be illicit. None were mentioned, possibly because competition for sales is coming from another source. It seems probable that opaque beers sold with new brand names (although often still referred to as Chibuku and available in more stylish containers) may be attracting the attention of many new customers. It has been suggested that another attraction might be a higher concentration of alcohol in these newer products. *Kachasu* always had a high alcohol concentration compared with brewed beverages and was probably drunk for that reason. It is impossible to say whether there was a public demand for higher concentrations of alcohol or whether they were supplied and a demand for more "mature" Chibuku grew out of this. The main lager, Mosi, has maintained a relatively low concentration of alcohol (4%) for many years; some of its clear beer rivals have higher concentrations. There can be no doubt that higher intakes of ethanol in a population generally may increase the harmful effects of drinking, and a strong argument can be made for controlling that trend.

Any research, even if only essentially descriptive, should raise more questions than it answers. This survey has done so with regard to methodology, the lifestyle of the drinkers who filled in diaries for us, and the implications of the findings for both the alcohol industry and the public health authorities.

REFERENCES

Allen, S. J., Meinzen-Derr, M., Kautzman, I., Zulu, S., Trask, U., Fideli Gao, F., & Haworth, A. (2003). Sexual behavior in discordant couples after HIV counseling and testing AIDS using biological markers in discordant couples from Lusaka, Zambia. *Lancet, 17,* 1–18.

Ambler, C. (1992). Alcohol and the control of labor on the copperbelt. In J. Crush & C. Ambler (Eds.), *Liquor and labor in Southern Africa* (pp. 339–366). Athens: Ohio University Press.

Babor, T. F., & Grant, M. (1989). From clinical research to secondary prevention: International collaboration in the development of the Alcohol Use Disorders Identification Test (AUDIT). *Alcohol, Health and Research World, 13*, 371–374.

Colson, E., & Scudder T. (1988). *For prayer and profit.* Stanford, CA: Stanford University Press.

Haworth, A. (1988). Alcohol problems among patients attending the University Teaching Hospital Casualty Department, Lusaka. *East African Journal of Medicine, 65*, 653–657.

Haworth, A., & Allen, S. (2001). Counselling couples as an HIV prevention strategy in PMTCT. Abstract. International Conference on Home-based Care, Chian, Mai, Thailand.

Haworth, A., & Chita, P. (2002). *A study of socially apart young people on the streets of Lusaka.* Report prepared for Healthlink International, London. Unpublished.

Haworth, A., Mwanalushi, M., & Todd, D. (1981). *Community response to alcohol-related problems in Zambia* (Community Health Research Reports 1–7). Lusaka: Community Health Research Unit, Institute for African Studies, University of Zambia.

Haworth, A., & Serpell, R. (1981). "Report presented to the Government of the Republic of Zambia, University of Zambia. Institute of African Studies" Lusaka, Zambia. (Unpublished)

Kay, G. (1960). *A social and economic study of Fort Rosebery* (Rhodes Livingstone Communications 21). Lusaka: Rhodes Livingstone Institute.

Lovelace, C. E. A., & Nyathi, C.B. (1977). Estimation of the fungal toxins, zearalenone, and aflatoxin, contaminating opaque maize beer in Zambia. *Journal of Food and Agricultural Sciences, 28*, 288–292.

Mukuka, L. (2000). *A baseline study of the extent to which illicit alcohol (Kachasu) is used and abused by periurban communities in Lusaka and Mazabuka.* Lusaka: Center for Social Policy Studies, University of Zambia.

Reilly, M., Nwegbe, M., & Ofafir, B. (1974). The methanol, ethanol and fusel-oil contents of some Zambian-alcohol drinks. *Medical Journal of Zambia, 8*, 13–14.

Chapter 5

Pilot Study on Patterns of Consumption of Nonindustrial Alcohol Beverages in Selected Sites, Dar es Salaam, Tanzania

Gad Paul Kilonzo, Nora Margaret Hogan, Jesse K. Mbwambo, Bertha Mamuya, and Kajiru Kilonzo

The United Republic of Tanzania is a developing country in East Africa. It covers an area 945,000 square kilometers and has 1,424 kilometers of marine coastline on the eastern border. It has long land borders with Kenya, Uganda, Rwanda, Burundi, Democratic Republic of Congo, Zambia, Malawi, and Mozambique, and it bestrides a number of inland waters. The wide variation in altitude offers a range of climates, from the humid and hot tropical climates on the coast to the warm savanna grasslands, the warm highlands, the temperate mountains, and the alpine climate on the slopes of the high mountains such as Kilimanjaro. It has a population of 34.6 million people and a population density of 39 per square kilometer (Tanzanian Bureau of Statistics, 2002). The country has about 120 ethnic groups; a common language, Kiswahili, facilitates easy communication between people.

There was a slow but steady increase in life expectancy at birth from 41.7 years in 1962 to a peak of 52 years in 1992, but this trend has been reversed over the past 10 years with a decline to a life expectancy of 48 years by 1998.

The AIDS epidemic is one major factor associated with this decline (UNAIDS, 2000).

Sustained per-capita income growth has been evident since 1995, with a steady increase from 0.6% per annum to an estimated 2.5% by 1999 (Bigsten & Danielson, 2001). Despite this overall increase, there are indications that income distribution has worsened over the years, and income has declined in absolute terms in the face of currency devaluation. It is estimated that 50% of the population live in poverty, most of them in rural areas (Mutalemwa, Noni, & Wangwe, 1998).

HISTORICAL PERSPECTIVE ON ALCOHOL AVAILABILITY AND CONSUMPTION

Tanzanians have brewed alcohol beverages from time immemorial. Homemade beers produced commercially by cottage industry have been licensed for over 50 years. The Intoxicating Liquors Act of 1968 provided detailed guidelines for their production and consumption. Licensing laws, however, have had little if any impact on production and consumption of homemade brews.

Cultural Practices

In traditional settings, patterns of production and consumption of alcohol were influenced primarily by cultural practices. Such occasions fell into the rhythm of cycles of cultivation and harvest together with life-cycle events, which exerted subtle controlling influence on drinking outside these ceremonial or social occasions. Pan (1975) emphasized that the accent was on "participation in the ritual rather than intoxication." In addition, well-established rules determined, for example, the appropriate strengths of alcohol beverages offered to various family members and age groups. The strongest drinks were reserved for the male elders attending the occasion. The strength of the drink was dependent on the period of fermentation prior to serving.

Ethnic and Geographic Areas

The varied climatic conditions, from the almost temperate highlands through the semiarid grasslands to the steaming hot coastline, influence economic activities and seasonal availability and type of alcohol beverages. Different peoples—the WaChagga, WaPare, WaBarbaig, WaIraqwi, WaMeru, WaArusha, and others—have various crops throughout the year that can be converted to alcohol beverages. The southern highland dwellers, for example, have *ulanzi*, which is a homemade wine tapped from a kind of bamboo called *mlanzi* available most of the year. The WaChagga and WaHaya have banana fruit that can be used to prepare beer. The WaPare mostly use sugar cane. Ethnic groups originating from the savanna grasslands depend more on the seasonal availability of grain harvests and the collection of wild honey.

In the past it was possible to discern an established alcohol production and drinking pattern among these ethnic groups, even when they were well established in urban areas. However, in the more recent past improved transport between regions of the country appears to have improved the availability of food crops from different climatic conditions in other regions of the country. The result may have reduced the differences between geographic locations in the production and consumption of alcohol. It is now possible, for example, for banana beer to be brewed throughout all regions of the country from bananas transported from growing areas in the highlands.

Production of Alcohol Beverages

Until the mid-1980s, production of industrially produced alcohol beverages had only managed to keep pace with population growth (Kilonzo, 1989). Between 1973 and 1983, inclusive, the availability (i.e., production plus imports minus exports) of alcohol beverages in Tanzania in terms of absolute alcohol (industrial beverages plus home brews) fluctuated around an average of 3.47 liters per person aged 15 and above (Kilonzo, 1989). Almost all available alcohol in the country is consumed. This figure places Tanzania among the countries in the medium consumption range, between Mexico, Kenya, and Gambia on the high side and Egypt, Morocco, and Sri Lanka on the low side among developing countries (Kortteinen, 1989). When records were examined in 1987, imported alcohol beverages amounted to less than 1% of available alcohol (Kilonzo, 1989). More recent records have not been examined; however, anecdotal evidence suggests that with trade liberalization and changing sociopolitical structures, industrially produced alcohol beverages occupy a more important position in certain strata of society today.

PATTERNS OF ALCOHOL CONSUMPTION WITHIN THE COUNTRY

Patterns of alcohol consumption in Tanzania appear to be changing under the influence of rapid urbanization and the breakdown of the traditional social fabric and family system. In urban areas that are ethnic melting pots, the prevailing social influences differ in many respects from the traditional setting. Peer influence is more likely to assume an inordinate importance as compared to the influence of elders or parents. Traditional customs and taboos are also likely to be less compelling in a heterogeneous ethnic confluence (Acuda, 1983). Market forces may also influence the reasons for brewing alcohol beverages and the settings in which they are consumed. Maula (1990) observed that the drastic fall in purchasing power of Tanzanian workers has influenced many people to get involved in the brewing of alcohol beverages as a way of supplementing their incomes.

In addition, the economic and sociopolitical changes noted earlier have had a negative impact on the formal (education) and informal structures for the socialization of young people, as well as on access to employment opportunities. Another noticeable change in pattern has emerged over the past 30 years or so in that alcohol beverages are increasingly taken for recreational use. In settings where there are few recreational facilities, drinking may be the main source of recreation. It has been suggested that these changes are accompanied by an increase in harmful patterns of alcohol use, including the use of alcohol beverages that are not produced industrially (Kilonzo, 1989; Singano, 1984).

A high proportion of Dar es Salaam residents appear to be nondrinkers of alcohol. In a survey of drinking behavior of persons aged 16 years and above in Dar es Salaam, 77.5% of male and 88.8% of female were reported abstainers (Ministry of Health, 1997). The 22.5% of drinking males comprised 6.5% who drank occasionally, 9.8% who drank on weekends, and 6.2% who drank daily. The proportion of females who drank was much less: 5.7% drank occasionally, 3.8% on weekends, and 1.7% daily. The Dar es Salaam region has relatively low drinking rates for Tanzania. The figures up-country are much higher (daily male drinkers were 14.7% in Morongo and 47.3% in Hai). Corresponding figures for females were 9% in Morogoro and 34.6% in Hai.

Among secondary school students in Dar es Salaam, regular use of alcohol varies from 4.8% occasionally (Musoke, 1997) to 6% at least once weekly (Kaaya, Kilonzo, Semboja, & Matowo, 1992; Kazaura et al., 1999). Among secondary school teachers in Dar es Salaam, 43.6% are teetotalers and 47.4% drink alcohol, out of which 15.8% drink at a hazardous level (Hussein, 1999).

Women are likely to be increasingly recruited into drinking as they brew traditional alcohol beverages for commercial purposes (Maula, 1988).

Types of Home-Brewed Alcohol Consumed

Studies have indicated that more than 89% of available alcohol in Tanzania is nonindustrially produced opaque beer (Kilonzo, 1989). The quantities and frequency of consumption of these beers are not well documented and have generally been difficult to study because of a number of problems, including the difficulty of estimating ethanol content and quantities consumed (Acuda, 1990). However, there are indications that a large proportion of the alcohol consumed in Dar es Salaam, especially among low-income earners, is in the form of home-brewed opaque beers.

The types of locally made/home-brewed alcohol beverages include the following:

- *Komoni*, a licit maize opaque beer made from maize husks.
- *Mbege*, a licit home-brewed beer made from banana and finger millet.
- *Mnazi*, a licit home-brewed wine made from palm juice.
- *Dengelua*, a licit home-brewed beer made from sugar cane and honey.
- *Wanzuki*, a licit home-brewed wine made from honey, mead, sugar, and yeast.
- *Gongo* (moonshine), an illicit spirit made from fermented papaw and sugar.
- *Kibuku*, an industrial opaque beer made from maize and sorghum.

Some studies have shown that some of these beers contain unacceptable amounts of aflatoxins, methanol, and other fusel oil (Mosha, Wangabo, & Mhinzi, 1996; Nikander et al., 1991; Wangabo, 1996). Anecdotal accounts continue to suggest that some impurities are deliberately added to the beverages to increase the intoxicating effects of the drinks. It is also generally observed that the drinking pattern in East Africa is that of bingeing (Partanen, 1991). Singano (1984) observed a "twice weekly" pattern of attending beer outlets.

Gongo

In addition to traditional home-brewed opaque beers, a proportion of the drinking population in Dar es Salaam drinks an illicitly distilled alcohol spirit called *gongo*. It is difficult to estimate the amount produced and consumed because of its legal prohibition. This form of alcohol, however, is much more toxic, sometimes containing methyl alcohol in concentrations that lead to blindness or even death. In gallows-like humor, "*gongo* drinkers" say to each other "*mazishi kesho saa nane*" (burial tomorrow at 2 p.m.), meaning that one of the drinkers might possibly die in a kind of Russian roulette. It is a challenge to estimate the available alcohol in this form and the proportion of the drinking population that consumes it.

During the 1960s efforts were made by the government to purify this illicit beverage by buying it from producers and distilling it industrially into *konyagi*. However, as the cost involved in the process was prohibitive, it was then produced from the direct fermentation of molasses.

RATIONALE FOR PILOT STUDY

A bold initiative set out in the Principles of Cooperation among the Beverage Alcohol Industry, Governments, Scientific Researchers, and the Public Health Community (O'Connor, 1997) involves producers of beverage alcohol, health professionals, and government representatives in a common enterprise of promoting healthy use of alcohol. The involvement of brewers of traditional opaque beers in this process cannot be considered without a better understanding of the consumption of these beverages. Traditional methods of collecting data have not given a very clear picture. The low reliability of self-reported measures in assessing alcohol use has implications for policy and management (Webb, Redman, Sanson-Fisher, & Gibberd, 1990). There is therefore a need to find better methods of studying the consumption of nonindustrial beverages.

METHODOLOGY

A method of keeping household diaries was piloted in July–September 2002 for the purpose of obtaining a clearer picture of frequency, quantities, settings, and social consequences of consumption of traditional home-brewed opaque beers. It was assumed that a prospective diary method with data recorded on a daily basis and monitored by trained research personnel would increase the reliability and validity of measurements of alcohol consumption, thereby minimizing problems encountered in quantity–frequency responses based on what participants remember (Webb et al., 1990) or choose to report.

Objectives of the Study

This pilot study had the objective of evaluating the diary method as an effective tool for obtaining information on the quantity and frequency of consumption of traditional home-brewed opaque alcohol beverages.

Description of Study Sites

The study was carried out at three sites—Gezaulole, Chamazi, and Kimara Baruti—near Dar es Salaam, the commercial capital of Tanzania, which has an estimated population of 3.5 million people. As a result of rapid rural–urban migration, it has grown from a coastal village with a population of 10,000 in 1894 to its present size. The majority of Dar es Salaam residents operate in the informal sector of the economy, earning their living through petty business or subsistence farming on the outskirts of the city.

Gezaulole village is situated 14 km to the southeast of the city. It has an estimated population of 5,000 people and is divided into three hamlets (Mbwa Maji, Mwera, and Kizani). Mbwa Maji is the oldest settlement of Swahili Zed coastal people who follow the Muslim culture. Consumption of alcohol, and of traditional beers, is expected to be low. Most of the inhabitants are peasant farmers, although a few are small-scale fishermen. Mwera hamlet is composed of a mixture of traditional inhabitants of the coastal area and migrants from up-country. Its population subsists entirely on peasant farming. Traditional beers and illicit spirits are produced and consumed. Some of the illicit spirits consumed in the city are thought to originate from this hamlet. Mostly migrants from up-country populate Kizani. The inhabitants subsist on small-scale farming; surplus tomatoes, sweet potatoes, and cassava are sent for marketing to the city center. A small proportion of the inhabitants are employed in the city and commute daily. Traditional beers, illicit spirits, and industrially produced beverage alcohol are consumed.

Chamazi village is situated 18 km southwest of the city, with a population of 6,000 people. Most of the population consists of traditional coastal inhabitants, with a small proportion of immigrants from up-country. Inhabitants subsist chiefly on agriculture, although a portion work in the city center, commuting daily. There are easy transport connections to the city center. Although traditional palm wine is the most popular drink, illicit spirit alcohol is also produced and consumed in the village.

Kimara Baruti is a village on the western side of Dar es Salaam about 14 km from the city center. The population is about 3,000. Most inhabitants are employed wage earners who supplement their income with small-scale mixed farming. They are relatively well-to-do. They could be expected to consume more industrially produced alcohol beverages than traditional beers. Illicit spirit is also available and consumed.

Sampling Method

Selection of the pilot sites was purposive, with the aim of capturing both areas mostly likely to consume traditional beers and communities of varying socioeconomic levels (except for the upper socioeconomic strata). From each village, three "ten-household" leaders were selected by lottery from the list of "ten-household" leaders kept by the village division or ward leader. All 10 households under the chosen "ten-household" leaders were recruited, making a total of 30 households for each village.

All the households and household members of the 10-cell units that were selected were requested to participate in the study. A total of 90 households (families) in the Dar es Salaam area were selected in that manner. The households averaged 3.2 members, of whom 2.21 were aged 12 years and over. A total of 199 household members aged 12 and over were therefore eligible to participate in the study.

Study Methods and Procedures

Personnel

A clinical psychologist, together with three social workers who trained and supervised 15 medical students as research assistants to conduct the study, assisted the principal investigator. The medical students were at the stage of clinical training and had experience in community fieldwork and research.

Three research teams were formed. Each team, consisting of one supervisor and five research assistants, was assigned to one of the three research sites (Gezaulole, Chamazi, and Kimara Baruti villages). An experienced research data entry clerk entered the data.

Method of Data Collection

Each member of the selected households aged 12 years and above was asked to keep a separate systematic daily record of the consumption of alcohol beverages over a specific time period. Diary records were not collected continuously but during specified periods of 4 or 5 days, which included midweek, midmonth, month-end, and weekend periods as well as a public holiday and other celebrations. In total the diaries were recorded at seven different points in time (five periods of 4 days and two periods of 5 days), with the cumulative total time for each household being 30 days (see Appendix 1).

The research assistants interviewed the head of the household about the drinking patterns of each member aged 12 years and above, and visited each household at specific intervals (at least twice during each 4-day period) (see Appendix 1) to review and monitor the diary recording with each household member. The research assistants worked with a maximum of five households at any one time. The drinking vessels were measured by volume. When a common drinking vessel was shared, the amount taken by each individual was estimated.

The research assistants also kept field notes on the diary record of each participant and noted discrepancies between the diary and information provided (during the initial interview) by the heads of household regarding the drinking behavior of household members. In addition, the research assistants made a daily summary of all their observations, including any obvious problems associated with drinking behavior.

Overall coordination of data collection was done by three trained supervisors. Supervisors performed reliability checks on the information being recorded and dealt with any procedural or research issues as they arose. The principal investigator and coinvestigators met with the research team to review progress and address issues arising. Information was collected on types and quantities of alcohol beverages being consumed and the amount of money used to purchase the beverages per day. Specimens of each type of traditional alcohol beverage were collected for chemical laboratory analysis. Information was also obtained regarding any member of the household who might be involved in brewing traditional beer, and the amount and income obtained from brewing.

Statistical Methods and Data Analysis

The data were precoded. The completed diaries were collected by the research assistants and checked by the research supervisors for reliability. The coded data were then entered into a computer for analysis using Statistical Program for Social Sciences 9.0 version for Microsoft Windows 1998.

Analysis of Alcohol Content

Specimens of some beverages were collected for analysis of alcohol content. From the three areas, one sample of each type of alcohol was collected from each of the three sites selected and analyzed for alcohol content and the possible presence of methanol or fusel oil. In a separate exercise, illicit alcohol distillate (*gongo*) was collected from the same sites of study in a similar way (March 2002). Four samples were taken from each site, covering weekdays and weekends. The results are outlined in chapter 11.

Equipment

A laboratory measuring flask was provided to every research assistant to measure the volume of drinking containers used by participants. Analysis of the alcohol beverages was done at the government chemist's laboratory.

Socioeconomic Analysis

Participants were grouped into three socioeconomic classes according to six weighted questions (see Appendix 2). The aggregate was then divided by three to get three socioeconomic status groups (lower, middle, and upper socioeconomic status). The range in points was 0–7 for the low socioeconomic class, 8–14 for the middle socioeconomic class, and 15–21 for the upper socioeconomic class.

Ethical Considerations

Consent from each head of household and each family member participating was requested and obtained. The research assistant was required to respect the confidentiality of each household member. The purpose of the study was explained to each family member that was involved in the study. After the collection of data was complete, a small token payment was made as compensation for loss of earnings during the period of participation in the study. Permission was sought and obtained from central and local authorities prior to embarking on the study.

RESULTS

The results presented cover the 30 days of data collection in Chamazi, Gezaulole, and Kimara Baruti. In total, 90 households participated in the study.

The number of household members ranged from 1 to 8, averaging 3.2. For those aged 12 and above the average number of participants per household was 2.21, yielding a total study sample of 199. All 199 study participants completed a diary. The majority of participants were adults between 16 and 45 years of age. Males (99) and females (100) were equally represented in the study. It is interesting to note that while male-headed households totaled 57 (63.3%), females headed as many as 33 (36.7%). Female-headed households did not reflect female dominance but rather single-female-parent families, widowhood, spinsterhood, divorce, and separation.

Out of the total 90 households selected, 89.5% of the household heads were teetotalers, but not all members of those households were teetotalers. Of the 199 study participants, 117 were nondrinkers and 82 were drinkers. All of the 90 household heads and members age 12 and above (drinkers and nondrinkers) were retained as study participants, as the primary objective was to evaluate the effectiveness of the diary method as an effective tool for gathering information. Retaining the nondrinkers allowed us to observe any change in drinking patterns during the study period (e.g., from nondrinker to drinker or vice versa). The data was used for comparative purposes with the WHO (World Health Organization) Alcohol Use Disorders Identification Test (AUDIT).

The first period of data collection, 29 July–1 August 2000 for Kimara Baruti and Chamazi and 30 July–2 August 2000 for Gezaulole (see Appendix 1), was considered a training period and was not used for analysis. It gave the research team an opportunity to address some of the initial problems encountered. Gender and religious affiliation did not appear to interfere with participants' readiness to complete the diary.

Alcohol Consumption

Seventy-nine (out of 82) identified drinkers specified the type of alcohol consumed during the period of recording. As predicted, the type of beverage consumed reflected the available food crop in the different study locations: *mzazi* (palm wine) at Chamazi and Gezaulole, *wanzuki* (honey mead) only at Chamazi, and *mbege* (banana beer) at all three study sites (Table 5.1).

The alcohol beverage most frequently consumed in the study population was beer, followed by *gongo,* an illicit spirit. When the data are disaggregated according to study site, *gongo* was consumed more in Gezaulole, the most rural of the three communities. Table 5.2 shows that beer and then *gongo* were the most frequently consumed alcohol beverages among both male and female drinkers.

When drinkers took more than one type of alcohol beverage, they would most likely be drinking beer and *gongo*. The results also show that *gongo* is the second most common alcohol beverage consumed by drinkers.

TABLE 5.1. Patterns of Alcohol Consumption at the Three Study Locations

Type of beverage	Chamazi Number T = 18	Gezaulole Number T = 42	Kimara Baruti Number T= 19
Illicit spirit			
Gongo	9	32	2
Licit traditional home-brew			
Komoni opaque beer	10	4	4
Mbege	4	10	2
Wanzuki	2	0	0
Licit industrial beverages			
Konyagi	3	0	1
Kibuku	8	3	2
Amarula	0	0	1
Beer	11	15	18
Licit palm wine home-brew			
Mnazi	8	14	0

Note. Beer refers to industrial beer. For homemade beers the specific names are used, e.g., *mbege*.

Alcohol consumption was higher among the older age groups in the study than the younger ages and this was significant ($\chi^2 = 0.00$). However, there is a fall-off among drinkers in the oldest age group from 65.4% (45–52 age group) to 48.1% (53+ age group) (Table 5.3). This is a common finding in many countries.

TABLE 5.2. Type of Alcohol Beverage Consumed by Gender

Type of alcohol beverage consumed	Female (34 drinkers) Number	%	Male (45 drinkers) Number	%	Total (79 drinkers) Number	%
Komoni	7	20.06	11	24.4	18	22.8
Kibuku	4	11.8	9	20.0	13	16.5
Mbege	7	20.6	9	20.0	16	20.3
Mnazi	8	23.5	14	31.1	22	27.8
Gongo	15	44.1	28	62.1	43	54.4
Beer	21	61.8	23	51.1	44	55.7
Wanzuki	1	2.9	1	2.2	2	2.5
Amarula	1	2.9	0	0.0	1	1.3
Konyagi	1	2.9	3	6.7	4	5.1
Total	65	39.8	98	60.2	163	

Note. Diary records reflect that drinkers consumed more than one type of alcohol beverage.

TABLE 5.3. Alcohol Consumption by Age

Age groups (years)	Drinking status		Total
	No	Yes	
12–20	30 (96.8%)	1 (3.2%)	31 (15.6%)
21–28	28 (70%)	12 (30%)	40 (20.1%)
29–36	24 (53.3%)	21 (46.7%)	45 (22.6%)
37–44	12 (40%)	18 (60%)	30 (15.1%)
45–52	09 (34.6%)	17 (65.4%)	26 (13.1%)
53+	14 (51.9%)	13 (48.1%)	27 (13.6%)
Total	117 (58.8%)	82 (41.2%)	199 (100%)

Socioeconomic Status of Drinkers

The study results indicated that different levels of socioeconomic status tend to consume different types of alcohol beverage (Table 5.4a). Drinkers from the upper socioeconomic group reported drinking beer and *kibuku,* both industrial alcohol beverages. Those from the middle socioeconomic group preferred to drink *gongo* and beer, whereas those from the lower socioeconomic group preferred *mnazi.*

There was equal representation between people of Muslim and Christian religious orientation in the study areas. Results in Table 5.4b rcveal that a greater majority of the Chrisitian population (48.7%) consumed alcohol compared to Muslims (31.0%). Chamazi, the study site with the greatest Islamic influence, was the only site where females consumed fewer units of alcohol than males.

TABLE 5.4a. Consumption of Alcohol Beverages by Socioeconomic Status

Type of beverage	Lower number	Middle number	Upper number	Total number
Komoni	2	16	0	18
Kibuku	1	10	2	13
Mbege	4	11	1	16
Mnazi	8	14	0	22
Gongo	2	40	1	43
Beer	5	37	2	44
Wanzuki	1	1	0	2
Amarula	0	1	0	1
Konyagi	0	3	1	4

Note. Diaries recorded consumption of more than one alcohol beverage.

TABLE 5.4b. Alcohol Consumption by Religious Orientation

Religious orientation	Percentage consuming alcohol		
	No	Yes	Total
Muslims	58	26	84
Christian	59	55	114
Traditional	1	0	1
Total	118	81	199

TABLE 5.5. Frequency of Consumption of Alcohol Beverages on 20 Weekdays

Number of weekdays on which alcohol beverage consumed	Chamazi (51 respondents)		Kimara Baruti (71 respondents)		Gezaulole (77 respondents)	
	Number	%	Number	%	Number	%
1 to 5	37	72.5	61	85.9	48	62.3
6 to 10	4	7.8	8	11.3	11	14.3
11 to 15	5	9.8	2	2.8	8	10.4
16 to 20	5	9.8	0	0.0	10	13.0

Frequency of Alcohol Consumption

The majority of drinkers at all three study sites consumed alcohol between 0 and 5 times during the week. Gezaulole, the most rural of the study communities, had the highest proportion (23.4%) consuming alcohol beverages between 11 and 20 times during weekdays (Table 5.5).

Diary records showed that 36.4% of Gezaulole drinkers consumed alcohol beverages four times or more over the weekends, compared to 23.5% and 12.7% in Chamazi and Kimara Baruti, respectively (Table 5.6).

TABLE 5.6. Frequency of Consumption of Alcohol Beverages During Weekends (10 Days, All Drinkers)

Number of weekdays on which alcohol beverage consumed	Chamazi (51 respondents)		Kimara Baruti (71 respondents)		Gezaulole (77 respondents)	
	Number	%	Number	%	Number	%
1 to 3	39	76.5	62	87.3	49	63.6
4 to 6	9	17.6	8	11.3	14	18.2
8 to 10	3	5.9	1	1.4	14	18.2

Quantities of Alcohol Beverages Consumed

The results summarized in Table 5.7 indicate that with the exception of Chamazi, female drinkers on the average consumed more units of alcohol (28.45) than male drinkers (26.31). However, women spent less money than men to buy their alcohol beverages. The amount of money drinkers reported spending ranged from 0 to 2.400 Tanzanian shillings (US$0–3) per day.

Proportion of Brewers by Gender

The study showed that 14% of participants were brewers. Males and females were equally represented, and production per day ranged from 16.5 to 22.3 liters. Earning averaged about US$4 per day.

TABLE 5.7. Mean Total Amount (Units of Alcohol) of Alcohol Beverage Consumed by Study Site and Gender

	Mean units consumed					
Study site	Period 2 (4 days)	Period 3 (4 days)	Period 4 (4 days)	Period 5 (5 days)	Period 6 (5 days)	Period 7 (4 days)
Chamazi (19 respondents)						
Females	13.83	13.66	25.66	40.66	42.16	42.0
Males	26.0	31.46	32.0	42.53	72.61	55.3
All	22.15	25.84	30.0	41.94	63.0	51.1
Kimara Baruti (22 respondents)						
Females	11.5	10.33	8.0	11.91	16.5	10.25
Males	13.2	9.7	7.0	6.9	6.2	4.8
All	12.27	10.04	7.54	9.63	11.81	7.77
Gezaulole (41 respondents)						
Females	34.7	38.58	25.29	45.64	46.88	33.11
Males	28.12	29.7	17.66	35.66	31.6	23.91
All	30.85	33.39	20.82	39.8	37.97	27.73

LIMITATIONS OF THE STUDY

The selection of study sites was purposive and aimed at capturing areas where traditional opaque beers were likely to be produced and consumed. The information generated thus cannot be generalized to the whole Dar es Salaam population.

Main Problems Encountered in Conducting the Study

A number of problems were encountered across sites, particularly during the first period of data collection. The research team needed to address those issues to ensure that the data being collected were reliable. For that reason, the data from the first period were not included in the analysis.

Problems encountered included the following:

Failure to keep scheduled appointments. This was a common experience encountered by all research assistants. Funerals, weddings, or prepresidential political campaigns involving the whole village often interfered with the research process.

Issues related to confidentiality. It was often difficult (although possible) to interview household members and review the diary privately. Some heads of households tended to be somewhat authoritarian and considered it their duty to safeguard the diaries of household members. In one case, the head of household had locked the diaries in a cupboard and in his absence the household members could neither complete their diaries as directed nor review them with the research assistant. Members of households who were unable to write relied on other family members to make diary entries for them.

Collecting data on illicit alcohol (gongo). Participants were concerned about the legal consequences of completing the diary. Police happened to be making arrests at the time the research was about to commence. Participants expressed fears that the diaries might be used as evidence and that the research assistants were in fact law informers. Political leaders reassured the people and the research continued.

Accuracy. One of the advantages of the diary method is the opportunity to review the record periodically and clarify any discrepancies. However, there was initially some concern about the accuracy of the recording. For example, one head of household (and his wife) had to make an unexpected journey, so he completed his own diary and those of three other household members in advance. Some diaries were incorrectly filled, and the participants were not always available for review. A few participants were reluctant to report on expenditure, in case their spouses saw the record. Other participants were suspicious and curious as to why a stranger should wish to know how much they had drunk and the expenditure they had incurred for alcohol.

Financial remuneration. Some participants indicated a reluctance to continue without some financial remuneration. Explanations regarding the ethics involved in "paying for information" appeared to be acceptable and they continued to participate.

Nonalcohol drinkers. Participants who did not consume alcohol felt it a fruitless exercise to continue to complete the diary after they had informed the research assistants that they were nondrinkers.

Number of visits. Some participants felt that there were too many visits to their homes to review the diaries. They suggested that although the diary method was acceptable, the review visits should be minimized.

During the first period of data collection, each of these issues was addressed and resolved as far as was possible, and the data collection then continued smoothly.

DISCUSSION

This pilot study was aimed at finding out if the diary method could yield reliable information regarding the types and quantities of nonindustrialized alcohol beverages consumed. A large proportion of Dar es Salaam heads of households appeared to be abstainers (89.5%), while 41.2% of all participants, including household members, consumed alcohol. This most likely reflected the fact that the household members were often young people. The results indicated that younger people were associated with higher levels of consumption of alcohol. They may also have reflected the choice of study sites (semiurban and rural rather than urban populations) on the periphery of the city, where traditional beers and illicit spirits are known to be brewed and consumed.

Types of Alcohol Beverages Consumed

The findings indicated that within the drinking population (41.2% of respondents) the majority concurrently consumed both industrialized and nonindustrialized alcohol beverages. Beer was the most common. Type of alcohol beverage consumed appeared to be associated with socioeconomic status. Drinkers in the upper socioeconomic status group reported drinking beer and *kibuku*, both industrial alcohol beverages. Those from the middle socioeconomic status group drank *gongo* and beer, whereas those from the lower socioeconomic status group drank *mnazi*. This is not surprising when one considers that an equally potent industrially made beer costs about five times more than a homemade beer. There also appears to be a rural–urban difference. In the most rural community, nonindustrial beverages (e.g., *gongo*) were more commonly consumed than industrialized beverages.

If study participants drank more than one type of alcohol beverage, they most likely drank beer and *gongo*. The results also showed that *gongo* was the second commonest alcohol beverage, consumed by both male and female participants in the study.

Frequency of Consumption

The majority of drinkers drank between 0 and 5 times during the week at all three study sites. The frequency of consumption was higher in Gezaulole, the most rural community. Gezaulolo had more participants who consumed alcohol beverages between 11 and 20 times during weekdays. Moreover, 36% of its participants consumed alcohol beverages more than 4 times during the weekends, compared to 23.5% and 12.7% in Chamazi and Kimara Baruti, respectively. Higher frequency of consumption appeared to be associated with more rural settings in this study.

With the exception of Chamazi (with its greater Islamic influence), female drinkers on average consumed more units (28.45) of alcohol beverages than males (26.31). However, females spent less money on alcohol than males.

It is also interesting to note that as many as 14.1% of the study participants admitted to being brewers, with virtually equal representation of males and females. The researchers gained the impression that this was an underestimate when correlated with information given informally by village leaders. Formerly, only older women brewed beer, and all women—including the brewers themselves—consumed only moderately. This may indicate that the brewing tradition has changed and that brewing has become just a business, for both men and women.

The findings indicated that traditional beverages, both licit and illicit, were popular with the majority of participants who consumed alcohol. Their availability and low cost compared to industrial drinks make them more attractive to low-income earners. Some monitoring and control of production, chemical contents, and consumption is essential. Effective strategies to achieve this may need to be considered.

Effectiveness of Diary Method

Previous research on the diary method indicates that its effectiveness in detecting heavy drinkers diminishes as alcohol consumption increases (Webb et al., 1990). This appears to be true for most methods of detecting heavy drinkers. The objective of this study was to evaluate the diary method as an effective tool for obtaining information on quantity and frequency of consumption of traditional home-brewed opaque alcohol beverages. It is not possible at this stage to do more than describe the findings on the quantity and frequency of consumption of traditional home-brewed opaque and other alcohol beverages as recorded by participants. There is no way of verifying the findings of this study. Neither can it be said that the diary method is more reliable and/or valid than other measures. The diary method would need to be administered concurrently with other data collection tools for comparative purposes.

In the current study, interviews with heads of households provided what can be termed collateral information, which may have some value as a countercheck for the diary record. However, the interviews with the household heads seem to have given them the impression that they had to exercise some control over the diaries of household members. Research supervisors and assistants addressed this issue during the first period of data collection and the issue seemed to resolve itself.

Other issues that were addressed during the first period of data collection were issues regarding confidentiality, the need for privacy, the importance of accuracy in recording, and the ethics of reimbursement for research data. Further problems encountered included participants' concern regarding disclosure of information for legal purposes, and complaints of the exercise being tiresome. After these issues were discussed with heads of households and household members there was improved cooperation.

The research team's evaluation indicates that apart from the first period of data collection, during which many problems were sorted out at all levels, the majority of participants found the diary method an acceptable way of recording patterns of alcohol consumption. The fact that the researchers had an opportunity to systematically review the recorded responses with individual participants meant that the completed diaries were likely to be more accurate than other self-report measures. It was noted that over the course of the month participants were increasingly accurate in recording and fewer discrepancies were found during the actual review of diaries.

From a research perspective this method is intensive and time-consuming, particularly in a culture where timekeeping is not an important cultural trait. In conclusion, the impression from this pilot experience was that it is a method that could be adapted for use in this cultural context, particularly where the data being collected are sensitive and have legal connotations.

APPENDIX 1: Study Time Periods, July–August 2000

The study time periods were as follows

Kimara Baruti and Chamazi

29 July–1 August
3 August–5 August
8 August–11 August
13 August–16 August
18 August–22 August
24 August–28 August
30 August–2 September

Gezaulole

30 July–2 August
4 August–7 August
9 August–12 August
14 August–17 August
19 August–23 August
25 August–29 August
31 August–3 September

APPENDIX 2: Estimate of Socioeconomic Status of Participants

1. What is your highest level of formal school education?
   No formal education (Score 0)
   Some primary (Circle) . . . Standard 1.2.3.4.5.6 (Score 1)
   Primary (completed standard 7) . . . (Score 2)

Some secondary (Circle) . . . Form IV (Score 3)
Secondary . . . Form V . . . VI (Score 4)
Technical school (Score 5)
College (Score 6)

2. What is the main roofing material of the house you now live in?
Grass, thatch (Score 1)
Palm leaves (Score 2)
Corrugated tin, cardboard, or plastic (Score 3)
Tin, tile, tar or with finished ceiling (Score 4)

3. Does your house have any of the following?
a. Pit toilet/latrine (Score 1)
b. Flush toilet (Score 1)
c. Tap drinking water (Score 1)
d. Electricity (Score 1)

4. Do you own the dwelling you now live in?
Yes (Score 2)
No, rented (Score 1)
Belongs to relatives/friends (Score 0)

5. What is your occupation? (Please write below)
No occupation (Score 0)
Occupation indicated (Score 1)

6. What is your monthly income in Tanzanian shillings per month (______)?
0–30,000 (Score 1)
30,000–80,000 (Score 2)
80,000–130,000(Score 3)
130,000 and above (Score 4)

REFERENCES

Acuda, S. W. (1983). Alcoholism, patterns of use, prevalence and prevention. *Proceedings of the 22nd Postgraduate Seminar, Kilimanjaro Christian Medical Centre,* Moshi, Tanzania, 3–6 May.

Acuda, S. W. (1990). Alcohol research in developing countries: Possibilities and limitations. In J. Maula, M. Lindblad, C. Tigerstedt, & L. Green-Rutanen (Eds.), *Alcohol in developing countries: Proceedings from a meeting in Oslo, Norway, 7–9 August 1988* (pp. 15–17). Helsinki: Nordic Council for Alcohol and Drug Research.

Bigsten, A., & Danielson, A. (2001). *Tanzania: Is the ugly duckling finally growing up? A report for the OECD project "Emerging Africa."* Uppsala: Nordiska Afrikainstitutet Research Report 120 (22–26).

Hussein, A. (1999). *Assessment of the prevalence of alcohol consumption, degree of dependence and associated factors among secondary school teachers in Dar es Salaam.* Community Medicine Field Project. Dar es Salaam: Muhimbili University College of Health Sciences (MUCHS).

Kaaya, S. F., Kilonzo, G. P., Semboja, A., & Matowo, A. (1992). Prevalence of substance abuse among secondary school students in Dar es Salaam. *Tanzania Medical Journal*, *7*, 21–23.

Kazaura, M. R., Kilonzo, G. P., Mbwambo, J. K., Kisesa, A. F., Aaro, L. E., & Klepp, K.-I. (1999). Alcohol consumption, psychophysical and emotional discomfort among school children: An experience from Tanzania. *Tanzania Medical Journal*, *10*, 30–34.

Kilonzo, G. P. (1989). Tanzania country profile on alcohol (WHO study on contribution of state monopoly systems to the control of alcohol-related problems). WHO Task Force, Stockholm, 7–14 September, 1986. In T. Kortteinen (Ed.), *State monopolies and alcohol prevention* (Monograph 181) (pp. 591–624). Helsinki: Social Research Institute of Alcohol Studies, University of Helsinki.

Kortteinen, T. (1989). Potential of state monopoly systems in preventing alcohol-related problems in developing countries. In T. Kortteinen (Ed.), *State monopolies and alcohol prevention* (pp. 23–53). Helsinki: Social Research Institute of Alcohol Studies, University of Helsinki. Report No. 181.

Maula, J. (1988). Research in the production and consumption of alcohol in Tanzania: Some preliminary reflections. In J. Maula, M. Lindblad, C. Tigerstedt, & L. Green-Rutanen (Eds.), *Alcohol in developing countries*. Helsinki, Finland: Nordic Council for Alcohol and Drug Research. Publication No. 18.

Maula, J. (1990). *Subsistence economy to cottage industry? Production of traditional alcoholic beverages as an income generating activity in Dar es Salaam region* (pp. 215–220). Licentiate thesis, Departmant of Sociology, Abo Academic University, Turku, Finland.

Ministry of Health. (1997). *Policy implications of adult morbidity and mortality—End of Phase One report* (pp. 98–99). Dar es Salaam: Ministry of Health.

Mosha, D., Wangabo, J., & Mhinzi, G. (1996). African traditional brews: How safe are they? *Food Chemistry*, *57*, 205–209.

Mutalemwa, D., Noni, P., & Wangwe, S. (1998). *Managing the transition from aid dependence: The case of Tanzania*. Paper presented at the African Economic Research Council/Oversees Development Council (AERC/ODC) Aid Dependence Workshop, December. Nairobi, Kenya.

Musoke, I. K. (1997). *Alcohol and drug abuse in Tanzanian schools: Experience from Dar es Salaam*. Research Report for Alcohol and Drug Information Centre (ADIC), Dar es Salaam, Tanzania.

Nikander, P., Seppala, T., Kilonzo, G. P., Huttunen, P., Saarinen, L., Kilima, E., & Pitkanen, T. (1991). Ingredients and contaminants of traditional alcoholic beverages in Tanzania. *Transactions of the Royal Society of Tropical Medicine and Hygiene*, *85*, 133–135.

O'Connor, J. (1997). *The Dublin principles*. Dublin: National College of Ireland.

Pan, L. (1975). *Alcohol in colonial Africa*. Uppsala: Scandinavian Institute of African Studies.

Partanen, J. (1991). *Sociability and intoxication: Alcohol and drinking in Kenya, Africa and the modern world*. Helsinki: Finnish Foundation for Alcohol Studies.

Singano, B. J. (1984). *The prevalence and patterns of alcohol consumption in urban communities of Dodoma municipality*. PhD dissertation, University of Dar es Salaam, Tanzania.

Tanzania Bureau of Statistics. (2002). *2002 Population and household census, general report*. Tanzania national web site, www.tanzania.go.tz.

UNAIDS. (2000). *Report on the global HIV/AIDS epidemic*. Geneva: Joint United Nations Program on HIV/AIDS.

Wangabo, J. L. (1996). *Investigation of the chemical constitution and impurities in traditional brews consumed in the Dar es Salaam area*. Master of science thesis, Department of Chemistry, University of Dar es Salaam, Tanzania.

Webb, G. R., Redman, S., Sanson-Fisher, R. W., & Gibberd, R. W. (1990). Comparison of a quantity–frequency method and a diary method of measuring alcohol consumption. *Journal of Studies on Alcohol*, *51*, 271–277.

Chapter 6

Licit and Illicit Beverages in Brazil

Magda Vaissman*

Brazil is the fifth largest country in the world and occupies half of South America. According to the 2000 Census, its population is around 160 million, with nearly 75% living in urban areas. It is a country of contrasts. The rich southeast region has a high concentration of economic resources, the best educational and health facilities, and the biggest cities including São Paulo and Rio de Janeiro. It is the wealthiest region of the country and is populated largely by descendants of the more recent European immigrants (including Italians, Poles, and Germans). On the other hand, the inhabitants of the northeast are generally descended from Portuguese colonizers, African slaves, and Native Indians. The northeast region, Brazil's poorest, has the country's highest rates of illiteracy and infant mortality. But although there is a great disparity of wealth, Brazil possesses many important assets that help to unite its population: the absence of significant ethnic, religious, or political conflicts; high rates of racial intermingling; and a single language spoken nationwide (Portuguese).

Brazil is officially a Roman Catholic country. However, Afro-Brazilian cults such as Candomblé and Umbanda are very strong in Brazil. The last 15 years have also seen a great expansion of Protestant and evangelical churches, which have started to exercise a strong cultural and political influence.

---

*Coordinator of Studies on Alcoholism at the Drug Users Assistance and Research Program of the Federal Institute of Psychiatry in Rio de Janeiro, and Coordinator of Post-graduation Courses on Chemical Dependence at the State University of Sá. Research team: Lia Ganc (Family Therapist); Luiz Alberto Matzbacher (System Analyst); Claudia Carvalho (Statistics).

HISTORY OF ALCOHOL BEVERAGE PRODUCTION IN BRAZIL

From Illegality to Licit Alcohol Beverage Production

Native Brazilian Indians were already acquainted with alcohol when the Portuguese first arrived in 1500. Numerous tribes produced *cauim,* a beverage made of corn or manioc and saliva. (Natives would spit into big vases, where the cereals fermented for days.) Production was on a small scale for ritual purposes, and it was unknown outside of native Brazilian populations (Carlini-Cotrim, 1997). The Portuguese brought with them port wine and *bagaçeira*, a distilled grappa beverage.

Today, the most popular distilled spirit in Brazil is *cachaça,* an alcohol beverage made from sugar cane. Discovered by chance between 1532 and 1548 (Almeida, 1997), primitive *cachaça* fermented naturally in a mixture of sugar cane residue and sugar cane juice that had been left outdoors in wooden plates and used for horse feed. It was soon offered as a beverage for African slaves. Later, it was distilled in special premises (*casa de cozer meis*) inside the sugar mills and was used as payment for African slaves (Almeida, 1997).

The Portuguese court, however, claiming that *cachaça* hampered production in the gold mines and interfered with the sale of Portuguese wines in the colony, then forbade its production, commercialization, and consumption. Furthermore, the courts decided to tax *cachaça,* and around 1756 it was onc of the most lucrative sources of funds for the reconstruction of Lisbon, which was destroyed by an earthquake in 1755. The *cachaça* taxes, known as subsidies, were also used to maintain the royal court (Santos, 2001).

As time passed, production techniques advanced and *cachaça* began to be better appreciated by the Portuguese, who consumed it at palace banquets and mixed it with ginger for religious festivals. This was the origin of the famous *quentão*. With the beginning of the coffee economy, final abolition of slavery in 1888, and establishment of the republic in 1889, all the things Brazilians widely cnjoyed became fashionable in Europe (Almeida, 1997).

In the 21st century, *cachaça* is no longer stigmatized as the drink of the poor masses. During the 20th century, the revival of the samba and the popularization of Carnival gave rise to *caipirinha*, an alcohol beverage composed of *cachaça* and lemon juice, which became famous worldwide for its taste and today is Brazil's national drink. *Cachaça* is now an international sign of high quality, appearing in fine restaurants and found in private cellars in Brazil and around the world (Santos, 2001). This beverage now appeals to a trendy, young, upper-class clientele. A variety of *cachaça* drinks can be ordered in any fashionable Brazilian bar or nightclub. Some of these drinks are the *caipitetra* (orange, honey, mint, and *cachaça*), the *granada brasileira* (passion fruit, cream, sugar, and *cachaça*), and the famous *caipirinha,* now an international Brazilian drink made of lemon juice, ice, and *cachaça. Cachaça* may soon have its own special day. The Brazilian Cachaça Society wants to declare 12 June International Cachaça Day. It was on this day in 1744 that the Portuguese prohibited its production and distribution.

Counterfeit Alcohol and Home-Brew Production

Cachaça is produced in both licit and illicit ways and is even subject to counterfeit production in Brazil. The state of Minas Gerais is the main Brazilian region of production and is also the leader in homemade production. Countless homemade products are commercially available for legal marketing with the payment of taxes for production and sale, although they are produced without any official registration, brand, or quality control. In most other Brazilian states *cachaça* is so easy to make that it is cheaper to buy than any other alcohol beverage (Carlini-Cotrim, 1997).

Although *cachaça* continues to be purchased by the lower classes, the prejudice against it is still present. Nevertheless, it continues its way to quality of purification, basically because of the potent combination of high alcohol content, cheap price, and ease of production in *alambiques*, home copper stills. The process followed is grinding of sugar cane, fermentation, distillation, and aging. Each producer claims to have a special secret that makes a unique and unforgettable drink, such as the type of cane, timing of the crop, duration of fermentation, addition of other ingredients, and wood used for the aging barrels.

The renewed popularity of the product has given rise to piracy, with many illicit versions being sold on the streets. For example, Havana, a popular licit brand made in the north of Minas Gerais, is considered to be one of the best *cachaças* in Brazil. A large part of Havana's appeal lies in the fact that it is very scarce because of the limited annual production of 600,000 liters. Havana's proprietor has declared that he will stop production simply because the many illicit varieties of *cachaça,* which do not pay the high taxes imposed on alcohol production, make it not worth producing his brand at all.

The Ministry of Agriculture technically controls the commercial production of *cachaça*. However, only about 10% of some 8,000 distilleries in Minas Gerais are registered. This poses massive problems for the standardization and commercial sale of the product. According to the Brazilian Spirits Association (ABRABE), Brazil has 25,000 brands of *cachaça*, and the country produces 1 billion liters of the beverage each year.

In terms of pure ethanol, *cachaça* ranks first, accounting for more than 85% of all distilled beverages marketed in the country. This represents 1.2–1.5 billion liters a year in recent years. Most of the production comes from two big companies. However, home and illicit production do not seem to decrease. The *cachaça* producers have not shared with other companies the "good times" mentioned earlier. Indeed, they have struggled to maintain sales at the same level over recent years, and have experienced a migration of consumers to beer or other kinds of spirits, although this phenomenon is present only in upper and middle classes.

Other Alcohol Beverages

Beer ranks first in consumption by volume. In 2000, 8.2 billion liters of beer were sold, according to the National Brewing Industry Syndicate (SINDCERV), with an estimated growth in production of 5% a year. Current annual per-capita consumption is about 52 liters, and SINDCERV predicts a rise to 65 liters over the next four years. In Brazil, the beer industry has created 130,000 jobs, with an aggregate value of some US$6.5 billion a year. It spends about US$200 million a year on marketing, and it sells at 1 million sites (SINDCERV, 2001).

The first brewery in the country, Bohemia, was established in Rio de Janeiro in 1853. Three decades later, Brahma and Antarctica, currently the two major breweries in Brazil, were established in the city of São Paulo. In 1998, these two companies formed a joint venture resulting in AMBEV, the company with the largest sales. As a result, AMBEV became the fourth largest company among global brewers, after Anheuser Busch, Interbrew, and Heineken. AMBEV holds more than 70% of the Brazilian beer market, with brands such as Antarctica, Brahma, and Skol; it also sells in Argentina and Venezuela. AMBEV bought a Uruguayan brewer, which owned Norteña and Prinz brands in early 2001, expanding its share of Uruguay's beer market to 48%. In addition, AMBEV sells Pepsi-Cola brands (including Lipton iced tea) and other beverages (mineral water, sports drinks). The company also brews Miller products locally.

There is no homemade production of beers in Brazil. Recent years have seen the introduction of some small fashionable artisanal, domestic breweries in important urban centers, but their output is negligible in the context of the overall Brazilian beer market.

Wine sales have been increasing sharply in recent years. From 1994 to 1995 alone, sales rose 35% because of large-scale imports from Europe, Argentina, and Chile. The national industry, consisting of small producers, is not accustomed to investing in marketing campaigns and competing with international and cheap wines. Four out of 10 bottles sold in Brazil come from abroad. Wine sales in Brazil are around 240 million liters a year, or 2 liters per capita. The importers are optimistic about increasing the sales still further (Carlini-Cotrim, 1997).

MARKETING AND LEGISLATION ON ALCOHOL IN BRAZIL

In July 1996, a federal law prohibited advertisements for alcohol beverages with an alcohol content above 13% on television between 6 and 10 p.m. However, beer may still be advertised without any restrictions as to time of day. Beer companies are the most important sponsor of sports such as football, music, artistic events, and the most popular party in Brazil, Carnival.

Alcohol beverages are widely available in Brazil because they are basically sold like any other industrial product, with few regulations and constraints. Alcohol beverages can be consumed and advertised on any day and at any time in parks, beaches, streets, restaurants, bars, bakeries, supermarkets, gas stations, factories, and hospital surroundings, and during sports events, funerals, and weddings. The main restrictions are: Alcohol beverages may not be sold to people under 18 years old, to persons with visible mental health problems, or to someone who is already intoxicated; and selling alcohol beverages on election days and driving with a blood alcohol content of 0.6 g/liter or greater are also prohibited. "Recurrent drunkenness" during working hours gives a company fair cause to fire an employee. Public drunkenness is also a legal offense (Carlini-Cotrim, 1997). Otherwise, there is no need for a specific license to sell alcohol beverages in ordinary shops or stores. They can also be sold near federal roads, although in São Paulo state, a specific law prohibits the sale of alcohol in establishments along highways. No special license is needed to sell *cachaça,* and there are no stores that sell alcohol beverages exclusively.

Environmental aspects seem to be one of the most important factors in studying *cachaça* consumption patterns in low-income communities. Such communities often present a gloomy landscape, with no leisure opportunities. To compensate, many corners have a *birosca* or *tendinha*—a very small shop, illegal in the slums—that sells alcohol beverages. Most of these commercial outlets are not legal and do not pay taxes. Often commercial establishments in the slums and other poor regions sell many products without a regular municipal license, and some homemade brands are sold as homemade alcohol products. Alcohol beverages are often sold in a similar way.

PATTERNS OF ALCOHOL USE IN BRAZIL

Cultural Variations in Alcohol Consumption

Brazil has very few scientific studies on alcohol use, and its drinking patterns have not been investigated as yet in community-based population surveys. Most studies have been carried out using an anthropological approach and qualitative methodology. Nevertheless, they cannot be generalized to Brazil's population as a whole because of its social, cultural, and economic disparities.

Santos (2001), studying a lower-class suburban community in Rio de Janeiro, showed that the drinking atmosphere at beaches, near the ocean, or in small lower-class bars (*biroscas* or *tendinhas*) is associated with leisure and social interchange, or provides the setting for a game of snooker, toto, or a soccer match on television.

However, alcohol intoxication is often observed in such locations, and the consumption of alcohol in this kind of public place is very high. Frequently, the intoxication has consequences such as fights and violence.

Guedes (2001) studied social relationships in bars and affirmed that these premises are a space of body exhibition and negotiation of "manliness," remembering that *machismo* is a central theme in the South American culture and is part of this imagine in social life.

As already suggested, the socioeconomic environment is an important factor in studying *cachaça* consumption patterns in low-income communities. In such communities, the *biroscas* or *tendinhas* often sell alcohol beverages illegally, escaping government taxes on what is sold. Because there is no supervision, there is no quality control of the products sold.

Popularly, *cachaça* and other distilled beverages such as cognac and liquors are termed "hot drinks," in comparison to "cold drinks" like beer. A worker may frequently be heard in a bar saying, "I'll take a hot one." In this kind of establishment beer is not sold; because there is no freezer or refrigerator, the only drink available would be a "hot" beverage and not a "cold" one.

Epidemiological Studies on Alcohol in Brazil

There are a few epidemiological studies that focus on patterns of drinking among various social groups in the general population in Brazil. These studies (Carlini-Contrim, 1997; Galduroz, Noto, Nappo, & Carlini, 2001; Galduroz, D'Almeida, Carvalho, & Carlini, 1999; Moreira, Fuchs, Moraes, & Breidmer, 2001; Notto, Nappo, Galduroz, & Mattei, 1993; Pechansky & BFC, 2001) suggest general patterns of use in urban populations. Drinking is a habit that starts early in life in Brazil, with abstention being relatively rare. Beer and *cachaça* are the most consumed beverages. Drinking for the sake of drunkenness does not seem to be the normal behavior in Brazilian society (Carlini-Cotrim, 1997). Cultural traditions and the strong presence of informal family controls over people's drinking seem to shape drinking behavior as a habit integrated into everyday life (Caetano, 1985).

There are occasions when drinking in excess does take place, as on New Year's Eve or during Carnival. Carnival, the most popular festival in Brazil, starts exactly 40 days before Easter, lasts for 3 days, and allows the rules and rites of everyday life to be turned upside down. This celebration is a period during which there is a break with daily norms, when people wear costumes, play, dance, and sing. In many cities men wear women's clothes, housewives dress up as prostitutes, and the poor array themselves as kings and queens. Rio de Janeiro hosts a samba parade, with beautiful seminude women dancing the samba—images that are broadcast all over the world. There is much drinking—mostly beer and *cachaça*—with some negative consequences (Oliveira & Melcop, 1997).

In 1999, the Centro Brasileiro de Informações sobre Drogas Psicotrópicas (CEBRID) conducted the first national household drug use survey in 24 cities in the state of São Paulo. It found a lifetime prevalence of 53.2% for alcohol consumption. That result was below the 83.6% rate found in Chile and above the 35.5% rate found in Colombia (all age groups). As in other countries, alcohol consumption is more common among males. Broken down by gender, 48.8% of men consumed alcohol versus 23.6% of women. In other countries, differences in gender rates were less marked: 87.3% of males versus 80% of females in Chile, and 78.8% of males and 86.6% of females in the United States (CONACE, 1996; Ospina, 1996; Packer, Davis, Krotil, Porter, Chromg, Virag, Gtroerer, Wright, Gustin, Greenblatt, Epstein, & Kopstein, 1999).

The prevalence of regular (3–4 times a week) alcohol consumption in São Paulo state were 8.0% for males and 1.2% females, equivalent to populations of 584,000 males and 89,000 females, respectively. The rates of alcohol dependence were 10.9% for males and 2.5% females, with a mean rate in the 24 big cities in São Paulo estimated at 6.6% (compared to Denver's rate of 4.5% and Atlanta's rate of 4.4% in the United States) (Branson, Porter, Guess, Witt, Virag, Gtoerer, & Gustin, 1996). The prevalence of symptoms of tolerance and other physical risks and alcohol-related problems was estimated at 3%.

FIELD RESEARCH

Characteristics of Jardim Metrópoles, São João de Meriti

Research was conducted daily among a group of families living at Jardim Metrópoles, a neighborhood of the municipal district of São João de Meriti, Rio de Janeiro state. Considered the most densely populated municipal district in Brazil, São João de Meriti has a population of 440,052 inhabitants occupying an area of 34,000 square kilometers. Jardim Metrópoles has a population of 26,000 and is part of Baixada Fluminense, a region on the outskirts of Rio de Janeiro. According to the Brazilian Institute of Geography and Statistics (IBGE), this is a low-income region of poor people with a monthly per household income below US$50 (the minimum monthly wage is around US$69).

The houses in the study area are made of cinderblock and brick, and water is provided for basic sanitation. The neighborhood has electricity and periodic public garbage collection. There are also poorer districts, where lower income families live in precarious housing conditions with no basic sanitation or drinking water. The population is typified economically by informal workers, who live on odd jobs such as carpentry.

Regarding preventive medicine, during the last 8 years the Brazilian Ministry of Health, in order to tackle the deep social inequalities, has enacted the Family Health Program (PSF). This seeks to introduce primary health care and has a mission to provide health promotion, treatment, and rehabilitation using a health team composed of general practitioners, nurses, and nurse assistants. The program works in parallel with the Health Agents Program (PACS), which conducts activities in health promotion, health education, and teaching the foundations for the gradual adoption of basic sanitation. In 1994, Brazil had 30,000 health providers; now there are 154,000 primary care agents, who cover 4,179 towns and 91 million inhabitants.

The study population receives periodic household visits from community health workers from the primary care program—the PSF—for the area, with the goal of improving the health status of the population. In the case of Jardim Metrópoles, PACS provides 13 community health workers coordinated by a nurse who specializes in primary care assistance. These community health workers assisted the investigators in selecting the participants for the study, working as field researchers applying the diary method.

METHODOLOGY

Study Participants

The study was carried out with the support of PROJAD, the Program on Alcohol and Drugs of the Institute of Psychiatry, Federal University of Rio de Janeiro. The project was approved by two ethics committees, one at the university and the other at the National Council of Ethics in Research of the Federal Ministry of Health.

For the proposed study, 139 people were selected in a nonrandomized sample and interviewed by the 13 community health workers who worked in the study area. These health workers administered the AUDIT questionnaire (Babor, de la Fuente, Saunders, & Grant, 1992) on 3, 4, and 5 December 2001 to one or more family members who consumed alcohol beverages. Each health worker was responsible for seven households. People who scored 8 or more were selected and introduced to the research protocol.

The research protocol consisted of explaining the basic responsibilities of the participants, including the completion of a daily journal on their drinking habits. They were asked to sign a consent form and to give basic information about themselves. If they accepted the research protocol and were aged 15 years or older they were included in the study. Of the 139 people interviewed by the health workers, 91 were selected from 77 families.

It is important to note that as the study sample was selected purposively (because the interviewees were not selected at random, nor was the sample size statistically based), the findings of the study cannot be generalized to the population as a whole in the absence of a probabilistic method.

Descriptive Analysis

During the research 10 respondents were lost to the study for various reasons, including travel and lack of interest. Thus, the results of the study were obtained through research effected with 81 people from the county of São João de Meriti, Rio de Janeiro, who were visited during a period of 6 nonconsecutive weeks (at 5 consecutive days per week) by health workers. Each weekly period presented different characteristics (middle of the week, end of the week, holidays, etc.). At each of the scheduled visits, a questionnaire on alcohol ingestion was completed. Thus, 30 questionnaires were completed for each respondent yielding information on types of beverage consumed, number of drinks consumed,* and so on. The description weekly periods in which the interviews were effected were as follows:

Week 1: middle of the week + weekend.
Week 2: Christmas.
Week 3: middle of the week.
Week 4: weekend + holiday (day off).
Week 5: end of the month.
Week 6: middle of the week before Carnival.

The main objective of the study was to compare the use of homemade *cachaça* and other types of alcohol beverages in a group of interviewed people. It is worth mentioning two points here:

- Because Brazil is an extremely hot country, it would not be expected that *cachaça* consumption would be very high. The dominant beverage is beer.
- One of the weeks studied included Christmas, a traditional celebration that is very important all over the country, when an increase in the consumption of alcohol beverages would be expected. Wine is more heavily consumed during this time of the year, even among those who do not usually drink alcohol beverages. Hence, the results presented may not correspond to reality during ordinary periods when there is no celebration.

Although the questionnaires filled in by the volunteers showed many variations, this chapter comments only on the basic findings. It is worth highlighting that questions related to money (i.e., personal income, family income, and amount spent on alcohol) posed problems: The majority of the respondents did not answer such questions, and those who answered did so with total imprecision.

It was not within the scope of the study to evaluate the existence of trends (increasing or decreasing) in the alcohol ingestion pattern with time, nor to consider the reasons for alcohol ingestion or its implications.

RESULTS

Table 6.1 shows the distribution of respondents according to their demographic characteristics. In other words, the 81 participants are mapped according to

*The amount of alcohol ingested is based on measures of one glass, which corresponds to 150 ml.

TABLE 6.1. Respondents by Sex, Ethnic Origin, Age Group, Level of Schooling, and Activity

		Schooling and activity group[a]														
		Illiterate		Incomplete elementary schooling			Completed elementary schooling		Incomplete junior high schooling		Completed junior high schooling		No answer			
Sex and ethnic origin	Age	1	3	1	2	3	1	3	1	3	1	2	1	2	Totals	
Male																
White	Less than 20 years										1				1	
	20–29 years			4							2				6	
	30–49 years	2		5			1				2		1	1	12	
	More than 49 years													2	2	
	No answer			1											1	22
Black	20–29 years			3							1				4	
	30–49 years			3			1								4	
	More than 49 years					1	1					1			3	12
	No answer			1											1	
Yellow	30–49 years						1								1	1
Brown	Less than 20 years			1		1									2	
	20–29 years			1											1	
	30–49	1		4		1	2	1		1					10	
	More than 49 years			1		1									2	15
																50

Female																
White	20–29 years						1								1	
	30–49 years			2		4	5	1							12	13
Black	20–29 years								1						1	
	30–49 years			1		3	1	1							6	7
Yellow	30–49 years						1								1	1
Brown	20–29 years			1											1	
	30–49 years			2		4									6	
	More than 49 years				1	1									2	9
Indigenous	More than 49 years					1									1	1
																31
Totals		1	2	30	1	17	14	3	1	1	6	1	1	3		
		3			48		17		2		7		4		81	

[a]Main activity groups: 1, work (study and work; work; housewife and work); 2, retired; 3, does not have own monthly income (seeking job; handicaped or disabled; housewife; does not study or work).

the variables recorded in the study, which were sex, ethnic origin, age, education, and basic activity groups.* Thus, the table provides the necessary information to draw a specific profile of the respondents, because no critical pattern of distribution was observed. What can be noted is the existence of many white males between 30 and 49 years old who work and have not completed their primary school education.

Table 6.2 shows the distribution of consumption of different types of alcohol beverages, according to the demographic characteristics of respondents. The table shows clearly that beer is the favorite drink of the groups interviewed, with the exception of the indigenous groups (only one such person, a woman, was interviewed), who prefer homemade *cachaça*. It is worth recalling that the consumption of homemade *cachaça* was much higher than of industrial *cachaça*.

The research showed that the highest consumption of alcohol beverages was in bars, *biroscas*, restaurants, and similar settings. The consumption of *cachaça* too was much greater in such places than in other settings. Beer consumption did not differ from one place to another.

According to the data presented in the tables, the consumption of alcohol beverages was not proportional to the number of people belonging to each category. Thus the following comments may be made about the group studied:

- Men drank more than women. However, the drinking of homemade *cachaça* was greater among women.
- The respondents below 49 years of age drank the greatest amount of alcohol beverages. However, people over 50 years of age drank more homemade *cachaça*.
- Although the people who drank the greatest quantity of alcohol beverages were the people who had incomplete primary school educations, illiterate people (who could not read or write) drank more homemade *cachaça*.
- The greatest consumers of alcohol beverages were black males under 20 years of age who worked and had an incomplete primary school education.
- The greatest consumers of homemade *cachaça* were black women over 49 years of age who either were retired or did not have their own monthly income, and did not read or write.

Table 6.3 presents the distribution of respondents according to the occasion and the company during the time they were drinking. The table shows

*For the variable "activity group," respondents were aggregated in clusters in order to help identify monthly income. Group 1 contains people who study and work, only work, or are housewives and also work; i.e., group 1 consists of economically active people and those who have their own income. Group 2 contains retired workers, who are inactive but have their own income. Group 3 brings together those who do not have their own monthly income for any reason: people who are looking for a job, are handicapped or disabled, are housewives, or otherwise do not study or work.

TABLE 6.2. Distribution of Consumption in Study Period by Age Group and Level of Schooling

	Age group				
Alcohol beverage	Less than 20 years	20–29 years	30–49 years	More than 49 years	No answer
Homemade *cachaça*	0	24	497	702	0
Industrial *cachaça*	0	0	53	70	0
Beer	1,518	4,818	16,432	1,758	599
Wine	75	190	671	144	50
Other	11	69	304	85	0
Total	1,604	5,101	17,957	2,759	649

	Schooling					
Alcohol beverage	Illiterate	Incomplete	Completed	Incomplete junior	Completed junior	No answer
Homemade *cachaça*	182	586	327	3	215	0
Industrial *cachaça*	26	69	8	0	10	10
Beer	481	16,821	4,190	523	1,970	1,140
Wine	78	708	145	10	65	124
Other	53	213	31	0	117	55
Total	820	18,397	4,611	536	2,377	1,329

TABLE 6.3. Consumption of Different Types of Drinks According to Company and Occasion

	Company							
Alcohol beverages on special occasion	Alone	Friends	Relatives	Boyfriend/ girlfriend	Family	Others	No answer	Total
Homemade *cachaça*	54	28	2	0	0	0	0	84
Industrial *cachaça*	0	10	0	0	0	0	0	10
Beer	322	3,737	3,127	97	1,604	132	0	9,019
Wine	45	352	188	12	156	0	0	753
Other	14	53	42	0	20	0	0	129
Total	435	4,180	3,359	109	1,780	132	0	9,995
Alcohol beverages on normal occasions								
Homemade *cachaça*	571	483	0	78	7	0	0	1,139
Industrial *cachaça*	48	65	0	0	0	0	0	113
Beer	2,366	8,818	2,320	1,308	1,280	12	3	16,106
Wine	147	159	45	6	20	0	0	377
Other	162	166	0	7	5	0	0	340
Total	3,294	9,691	2,365	1,399	1,312	12	3	18,075

that the highest consumption occurred on occasions that were not particularly special, which means that there was no need for special motivation in order to drink. As to company, the highest consumption occurred among friends.

It can be seen that on special occasions, alcohol beverages were consumed particularly among family, friends, and relatives. During special events, some respondents also drank alone. It is important to note that the consumption of *cachaça* jumped on occasions that were not special. The consumption of beer was consistently highest on whatever occasion (special or not) and whether or not in the company of friends.

CONCLUSIONS

The results presented show that homemade *cachaça* (whether illicit or not) is one of the population's favorite drinks in the area of São João de Meriti, second only to beer. It was also shown that, in general, the study population needed no special occasion, company, or place in order to drink. Consumption of alcohol beverages was highly independent of sex, age, color, and other characteristics reviewed. However, it should be emphasized again that the survey was small and not random. The results could have been strongly influenced by an outlier* and might not correspond to the reality of the population of São João de Meriti as a whole.

A further study could well be effected with other and larger samples of the Brazilian population, in order to gain a more accurate picture of patterns of alcohol beverage consumption in Brazil, in light of differing cultures and social classes.

REFERENCES

Almeida, R. (1997). *História da Cachaça. Beco da Cachaça.* Associação Mineira de Aguardente de Boa Qualidade (http://www.cachaca.com/baronesa.htm).

Babor, T. F., de la Fuente, J. R., Saunders, J. B., & Grant, M. (1992). *AUDIT: The Alcohol Use Disorders Identification Test: Guidelines for use in primary health care.* Geneva: World Health Organization.

Branson, S., Porter, B., Guess, L., Witt, M., Virag, T., Gtroerer, J., & Gustin, J. (1996). *National household survey on drug abuse: Population estimates 1995.* Rockville, MD: Substance Abuse and Mental Health Administration, Office of Applied Studies.

Caetano, R. (1985). Alcohol-related observations in Rio de Janeiro. *Drinking and Drug Practices Surveyor, 20*, 7–10.

Carlini-Cotrim, B. (1997). *Alcohol use and related problems in Brazil: A review. Alcohol and related problems around the world.* Geneva: World Health Organization.

CONACE. (1996). *Consejo Nacional para el Control de Estupefacientes, Ministério del*

*In this context, an "outlier" is someone who presents discrepant alcohol ingestion in relation to the rest, i.e., information that "escapes" the pattern.

Interior Segundo Informe Anual sobre la Situación de Drogas en Chile (p. 348). Chile: Sistema Nacional de Información sobre Drogas.

Galduróz, J. C., D'Almeida, V., Carvalho, V., & Carlini, E. A. (1999). *III Levantamento sobre o uso de drogas entre estudantes de 1º e 2º grau em 10 capitais brasileiras.* São Paolo, Brazil: CEBRID, Centro Brasileiro de Informações sobre Drogas Psicotrópicas, Departamento de Psicobiologia da Escola Paulista de Medicina.

Galduróz, J. C., Noto, A. R., Nappo, S. A., & Carlini, E. A. (2001). *I Levantamento Domiciliar Nacional sobre o uso de drogas psicotrópicas.* São Paulo: Universidade Federal de Medicina de São Paulo, EPM, Departamento de Psicobiologia (FAPESP).

Guedes, S. L. (2001). *EDUFF, Jogo de Corpo.* Niteroi.

Moreira, L. B., Fuchs, F. D., Moraes, R. S., & Breidmer. (2001). *Alcoholic beverages consumption and associated factors in Porto Alegre, a southern Brazilian City: A population based survey.* Unpublished manuscript.

Notto, A. R., Nappo, S., Galduroz, J. C., & Mattei, R. C. E. A. (1993). *III Levantamento sobre o uso de drogas entre meninos e meninas em situação de rua decinco capitais brasileiras.* São Paolo: CEBRID, Centro Brasileiro de Informações sobre Drogas Psicotrópicas, Departamento de Psicobiologia da Escola Paulista de Medicina.

Oliveira, E. M., & Melcop, A. G. (1997). *Álcool e Trânsito* (p. 120). Recife: Instituto Raid. CONFEN-DETRAN-PE.

Ospina, E. R. (1996). *Estudio nacional sobre consumo de substancias psicoactivas.* Bogotá, Colombia: Fundación Santa Fé.

Packer, L., Davis, T., Kroutil, L., Porter, B., Chromg, J., Virag, T., Gtroerer, J., Wright, D., Gustin, J., Greenblatt, J., Epstein, J., & Kopstein, A. (1999). *Summary of findings from the 1998 National Household Survey on Drug Abuse.* NHSDA series: H-10. Rockville, MD: Substance Abuse Mental Health Administration, Office of Applied Statistics.

Pechansky, F., & BFC. (2001). Problems related to alcohol consumption by adolescents living in the city of Porto Alegre. *Journal of Drug Issues, 25*(4):735–750.

Santos, J. J. (2001). *Cachaça não é água* [Cachaça is not water.] Monograph submitted for diploma in chemical dependency, Estacio de Sá University, Rio de Janeiro.

SINDCERV. (2001). *Cerveja com responsabilidade e só alegria.* São Paulo: Sindicato Nacional da Indústria da Cerveja.

Chapter 7

The Reporting of Alcohol Use Through Personal Diaries in Two Mexican Communities

Haydée Rosovsky

MEXICO: A COUNTRY WITH A MEDIUM LEVEL OF DEVELOPMENT

With an area of almost 2 million square kilometers, Mexico presents a variety of climates and landscapes. Although it is located in North America and has the United States—the most powerful nation in the world—on its northern border, its neighbors to the south, the Latin American countries, are closest to its historical, economical, and cultural identity. In 2000, there were 98 million inhabitants and, according to recent government data, 62.5% of them are poor (SEDESOL, 2002).

An increasing proportion of people live in urban centers and more than 75% live in large urban settlements, although many of them lack basic services and resources. There is continuous migration to these areas by indigenous Mexicans or peasants in search of some way of surviving; thousands risk their lives each year trying to cross the northern border, where they may earn less than the minimum legal wage but enough to obtain food back at home (Instituto Nacional de Estadísticas, Geografía e Informática [INEGI], 2002). Nevertheless, Mexico is not considered a poor country: According to macroeconomic global indicators, it is the world's ninth largest (Presidency of the Republic, 2002). But, as in other Latin American countries, the general growth and size of the economy and the movement of capital and investments have not secured a more balanced distribution of wealth. The gross national product (GNP) and other indicators say nothing about the quality of life of the people. According the United Nations Development Program (UNDP), Mexico is 54th, heading the medium-ranked countries in the Human Development Index, and followed by Cuba, Belarus, Panama, and Belize (UNDP, 2002).

Mexico is still a country of young people, although there have been some changes in recent decades: The proportion of the population under the age of 15 years was 50% during the first half of the 20th century, but by 1990 it was 40%, and in 2000 it was 37%. The demographic changes are dramatic as a result, among other reasons, of a reduction in infant mortality, longer life expectancy (in 2000, 71 years for males and 77 years for females) and fewer children born to educated mothers.

There are some improvements in other social indicators: Although in 1990 the illiteracy rate for people aged 15 years and over was 12.4%, in 2000 it was 9.5%. The proportion of children between 6 and 14 years old who did not attend school was 14% in 1990 and 8.2% in 2000. The average number of years of education for people aged 15 years and over was 7.6 in 2000. In 1990 only 9.4% of those aged 18 and over had an education beyond secondary school; by 2000, that proportion had grown to 12%.

Although most (87%) of the population is Roman Catholic, other religions have been gaining in importance during recent decades, such as Protestants and evangelical Christians, mainly among peasants and the working class.

According to the most recent census, in 2000, only half the population aged 15 and over are economically active and receive some income. Sixteen percent earn less than the minimum wage (US$127 per month); 30% earn between once and twice the minimum wage; 31.5% between two and five times the minimum wage; and 10.5% more than five times the minimum wage. Wealth is therefore concentrated in a small percentage of the population. The remaining half of the population consists of people who work but receive no money and are referred to as "self-consumers" (INEGI, 2002).

Mexico is a country that has experienced rapid modernization since the beginning of the 20th century. Nevertheless, its long and rich past and culture continue to be manifest. Mexico stands partly in the present, in modernity, but it is very much caught in the past, in old traditions and cultures, which, against all odds, still persist. This has to do with how the country has evolved and how the people who were there thousands of years ago have managed to survive through their legacy, their culture, their art and monuments, their food, and their traditions. Indeed, when they were conquered by the Spanish Crown and the Roman Catholic priests in the 16th century, they were subjected to another, more relevant, conquest: the *mestizaje*. They were mixed with the White men, and most of the population—chiefly the lower social classes—are descendants of that blending.

Mexican culture has the power of transforming or mixing the old with the new so that the conqueror becomes the conquered; Spanish food is enriched by Mexican fruits and vegetables, and the language is a mixture of words and expressions that is not the same as the language of Madrid. The Roman Catholic priest has to tolerate the practice of paganism in many local ceremonies and the celebration of pre-Hispanic religious festivities as part of the religious life of the communities, as happened during the Pope's recent visit to Mexico. This does not mean that past and present mingle harmoniously: There have been many clashes between tradition and modernity, as in recent years, with the Chiapas insurgency followed by much debate on the situation of the Indian communities and their legal rights.

This *mestizaje* or mixture of ethnic groups has also influenced the culture and the way Mexican society has evolved. Another important shaping force in the 20th century has been the strong economic and cultural influence of the United States, causing conflicts on a number of fronts.

HISTORY OF DRINKING AND ALCOHOL PRODUCTION

Pre-Hispanic Times

Alcohol production and use is an ancient practice in Mexico as well as in various parts of Central America. Long before the end of the 15th century, the peoples originally inhabiting the territory occupied by Mexico today produced and consumed alcohol beverages, with the main one being a fermented beverage called *pulque.*

Pulque or *octli* (original name of the beverage in the Nahuatl tongue) is a true survivor of the pre-Hispanic indigenous past. It results from the fermentation of *aguamiel*, or maguey (*agave*) juice, when the plant is at least 5 to 7 years old. *Aguamiel* is collected daily in the morning and afternoon using an *acocote*—a kind of gourd. The maguey stem, leaves, and thorns were used in daily life for many things such as shoes, clothing, and construction. *Aguamiel* contains 7–14% sugars; after fermentation, these decrease to 0.5–2% while its calorie content increases to 3–6.5%. The ethanol content by volume is similar to that of beer, at between 3 and 6%. Given its ethanol and carbohydrate contents, it is a good energy source; it is also an important source of vitamins and free amino acids.

Drinking in pre-Hispanic times, as well as the use of other mood-altering substances found in plants, had specific religious, spiritual, and healing meanings and purposes, which played an important role in structuring lives and beliefs. Eating *peyote* or drinking *pulque* had a mystic ritualistic character, with clear restrictions on who was allowed to use them or become intoxicated (Rosovsky & Romero, 1996; Viesca, 2001). Indians drank *pulque* at ceremonies and festivities under strict control, and those who drank it outside the established norms were severely punished. This was especially so for individuals of high social status, such as noblemen and priests.

Certain people were viewed as victims of a fixed destiny to become intoxicated with alcohol because they were born under an astrological sign, the "two rabbits day"; nevertheless, they were still punished, "because this *octly* and this drunkenness is the source of all discord" (Sahagun, 1969). *Pulque* was perceived as a divine substance that should only be consumed during rituals in ceremonies associated with agriculture, religion, and life-cycle events such as birth, marriage, and death (Taylor, 1979). According to historical evidence, alcohol use did not represent a serious problem in those days, perhaps because there was a balance and a clear division between strict control measures and tolerance for intoxication at ceremonies and celebrations.

Pulque was consumed not only for its intoxicating effects but also as medicine for the sick, nursing mothers, and the elderly; even today, it is consumed for its supposed medicinal properties, to correct digestive disorders, anorexia, asthenia, kidney infections, diminished lactation, and diarrhea. Different varieties of *pulque agave* grow practically throughout Mexico, but the main areas of cultivation are in the central states of México, Querétaro, Michoacán, Tlaxcala, and Hidalgo; the border between the last states is the region where the farming of *agave* is most intensive (Soberón, 1998).

Although not as important as *pulque,* other fermented alcohol beverages were made in pre-Hispanic times from maize, other types of maguey, and *peyote*, particularly in the north of the country among the Yaqui and Tarahumara Indians (Berruecos, 1983). Many of these old fermented drinks are still consumed today at local village markets in different regions of the country, some made with fruits (e.g., *capulín* wine and *tepache*), bark (*balché*), sap (*pulque* and palm wine), and roots, pulp, seeds, and stems (from cassava, maguey, *tesgüino*, etc.) (Vargas et al., 1998).

With the conquest and colonization by the Spanish empire, distillation from sugar cane started, and soon there were many types of *aguardiente* available for popular consumption. The conqueror Hernán Cortés developed the first alcohol sugar cane distillery in Cuernavaca, in the state of Morelos, as early as the 16th century (Lozano Armendares, 1998; Rosovsky, 1985).

Chinguirito—the popular name for sugar cane *aguardiente*—was first produced through a simple and cheap method. Water and syrup were blended in cow skins and placed near the stove to accelerate fermentation. The liquid was then poured into the still to make cane distillate. Although the Spaniards prohibited it during their almost three centuries of rule, mainly to protect Spanish trade, Indians enjoyed it on account of their marginal situation regarding the Crown's administrative control. Subsequently, as has been pointed out, "many people who lack other means of survival have devoted themselves to the production of prohibited beverages, especially sugar cane *aguardiente*" (Corcuera de Mancera, 1996). Corruption among minor officials and tolerance of such activities allowed them to continue. Finally, during the second half of the 18th century, for economic reasons, the authorities decided to allow the production and sale of sugar cane *aguardiente* (Lozano Armendares, 1998).

The collapse of Indian dominance and the eradication of the old order meant that the many social controls over alcohol were abraded and alcohol use became more problematic for the Indians, who presented high rates of drunkenness and social disruption. The loss of their freedom and social status, coupled with their growing poverty and feeling of humiliation, were surely important influences on their drinking habits. *Pulque* became socially rejected as a drink by the Spaniards and the Creoles (Corcuera de Mancera, 1996; Vargas, 1999).

The Spaniards also introduced grape growing for wine making, and alcohol production became an important source of wealth for the Spaniards and Creoles, and later on for some mestizos. As Bunzel (1940) reported, Indian peasants were paid with *pulque* or alcohol by the coffee plantation owners in Chiapas and on other big ranches, a practice that continued until recent decades. Alcohol thus became another tool of domination.

Traditional Distilled Beverages

Most traditional distilled beverages are *aguardientes* (made with sugar cane, grapes, or other fruits) and the rest are *mescals*. Various beverages made from different plants are found in specific regions of the country, such as *bacanora* (*agave*), *charanda* (sugar cane), *colonche* (red prickly pear), *posh* (grape), and *sotol* (maguey). The autochthonous traditional distilled beverage that is more widely known internationally is *tequila*, which has become a symbol of the Mexican multiethnic culture (Lozano Armendares, 1998; Vargas, 1998). *Tequila* is obtained from the distillation and refinement of musts prepared from macerating the maguey *Agave tequilana* Weber, blue variety. It is permitted to add up to 49% of sugars. *Tequila* must be allowed to age in oak barrels until a yellowish crystalline color is obtained, and it must only be made with the right *agave*s, which grow in the states of Jalisco, Guanajuato, Michoacán, Nayarit, and Tamaulipas.

Mescal is obtained from the distillation and refinement of musts from different types of *agave*s. Either all sugars come from the *agave*s (100% *mescal*), or up to 20% of other sugars may be added. According to its denomination of origin, the raw material should come from the states of Guerrero, Oaxaca, Durango, San Luis Potosí, and Zacatecas. Traditional beverages such as *tequila, mescal,* and, more recently, *sotol* have acquired a "denomination of origin" at the World Organization of Industrial Ownership to guarantee their authenticity and to ensure that only Mexican raw materials are used in their production.

Independence and Industrialization

After independence from Spain in 1821, and as industrial development progressed, the conditions were present for the production and consumption of various other beverages on a large scale. The success of beer, which later became the main industrial alcohol beverage consumed in Mexico, was tied from its beginnings to the development of other important industries, such as glass, refrigeration, cardboard packaging, and transportation. Its manufacture was initiated in the middle of the 19th century, and today the industry is one of the strongest in the world. Industrial production of *tequila* also dates from this time; both these industries started as family businesses.

Industrialization of the country, which was accelerated after the Second World War, had important implications for shaping the alcohol industry and alcohol drinking patterns. Alcohol industries, like other manufactures, shifted from traditional producing countries to new markets with the introduction of new Western beverages. In the early 1950s, enormous amounts of foreign capital were invested in Mexico, and new, costly marketing techniques were used to promote alcohol beverages and other goods. Several multinational alcohol producers and advertising agencies became established in Mexico. At that time the government encouraged the development of the alcohol and wine industries through fiscal stimuli and other actions. For instance, the state of Aguascalientes was considered to have suitable climatic conditions for grape wine growing, and, together with Coahuila and Baja California, it became one of the country's wine-producing areas.

Increasing urbanization and the influence of foreign lifestyles through new means of communication such as films and television also had an impact on drinking patterns and played an important role in making people want to consume foreign goods and behave differently, especially in urban areas. All these phenomena were taking place in conjunction with other important economic, social, and cultural influences that were promoting Western lifestyles in the postwar social climate.

Current Situation

The range of industrial beverage production in Mexico includes both fermented beverages, such as beer and wine, and distilled spirits, which offer the greatest variety of products. After beer, distilled beverages have the highest presence in the commercial market. The most common distilled drinks in Mexico are brandy, rum, *tequila*, and more recently vodka; other beverages that follow in importance are fermented wines, liqueurs, mixed drinks, and cocktails, such as coolers.

Mexico has a number of legal measures to regulate different aspects of alcohol production, trade, distribution, and consumption. Various official standards specify the requirements for the preparation of beverages, in terms of the properties of each beverage, health considerations, and other characteristics such as the type of containers and labeling. There are some gaps in the legal framework, as well as lack of proper enforcement of the law.

As mentioned, some traditional beverages have become industrialized as part of the commercial market and are distributed throughout the country, with growing exports. Among these, beer and *tequila* are outstanding examples. More recently, *mescal* and *sotol* have started to compete in the local and international commercial markets, trying to emulate the success of *tequila* in foreign trade.

Other traditional alcohols—fermented or distilled—are produced on a smaller scale, even in a domestic and/or clandestine fashion, and are distributed and sold locally in a quite informal way. These include *pulque* and *aguardiente* beverages, which lack most of the already mentioned controls and where no records of their production and sales are kept. Peasants, Indians, or other local people from low-income groups drink them. Potable alcohol of 96% content by volume can be included in this category. It is consumed in urban and rural areas by the poorest people, sold in small unbranded bottles in small grocery stores, drug stores, or markets. *Pulque* is also sold in Mexico City in working-class neighborhoods in special establishments called *pulquerías*, places full of tradition with musicians and typical decoration. In addition, there is a counterfeit market for other beverages that either are stolen or are imitations of commercial/industrial products.

Mexico has serious problems with enforcement of the law, with respect not only to alcohol but also to many other goods that are sold freely, such as toys, recorded music, videos, electronics, and clothing. Many of these products are imported illegally into the country; others are stolen or made in Mexico in imitation of international brands and then sold in the informal market, on the streets, without paying taxes. This situation is harmful for the government and society at large. The authorities seem unable to control the problem and there is a huge demand for this unlawful market, because of its low prices.

Alcohol Drinking Patterns

Drinking alcohol has an important role in social interactions in all population groups in Mexico. For instance, a native of Chiapas must offer or exchange a "drink" (*trago*) when facing authority, encountering friends, sealing a business deal, or asking a favor. In urban areas, alcohol accompanies meals, parties, and occasions for happiness and sadness, as is shown by this common phrase: "For all that is bad, *mescal,* and for all that is good, too." Drinking is also relevant in sacred settings in a predominately Roman Catholic society. Nevertheless, the growing influence of other Christian religions, embraced mainly by Indians, peasants, and working-class people, is promoting abstinence among their followers, or at least the avoidance of distilled "strong" alcohol.

Though there are specific places where some type of beverages can be found, such as *pulquerías* (bars where *pulque* is sold) or *tepacherías* (which sell *tepache*), drinking of other, Western-type beverages can take place in a wide variety of settings such as saloons, coffee shops, restaurants, hotels, bars, and discos. Every social, civic, and religious event in Mexico includes drinking alcohol.

Since the 1970s, several epidemiological surveys have been carried out among the general population that allow patterns and trends of alcohol use to be identified. Another indicator for the estimation of alcohol intake is per capita consumption. Per capita alcohol consumption is based on the national sales of industrial beverages, so other types of beverage are not included. Alcohol per capita consumption in the Mexican population is estimated at around 5 liters of pure ethanol for people aged 15 and over. This amount is lower than the figures reported by other countries; for example, it is four times lower than in Spain, three times lower than in the United States, and almost half of the figure for Chile. But the estimation of per capita alcohol consumption has limitations, because it does not include nonindustrial beverages. Moreover, it presents a homogeneous picture of alcohol use, without providing information on the distribution by regions or gender (Rosovsky, 2001).

Alcohol sold in the country is not evenly consumed among the population, as was demonstrated by the three national household surveys on addictions carried out between 1988 and 1998 among the urban population between 12 and 65 years old. Abstention rates (people reporting not drinking during the last 12 months) among males remained relatively stable between 1988 and 1998 (23% and 27%), but a significant change was observed among women, with fewer abstainers in the latter year (63.5% in 1988, 55.3% in 1998). The survey showed that the top 25% of heaviest drinkers were responsible for the intake of 78% of the alcohol consumed.

In the 1998 survey, among those who drank, 67.6% of males and 28.9% of females reported consuming beer; 41% of men and 23.3% of women said that they drank distilled beverages; 12.8% and 6.6% respectively drank coolers and other bottled cocktails; and table wine intake was reported by 11.6% and 6.6%, whereas *pulque* was consumed by 5.4% of males and 1.5% of females. *Aguardiente* and 96% pure alcohol were drunk only by 2.3% and 0.3% of men and women, respectively. Beer accounted for 50% of the pure ethanol intake of the urban population studied, distilled beverages for 32%, table wines 5%, *pulque* 9%, and *aguardiente* and 96% pure alcohol for 4%.

It is important to remember that these national surveys were carried out on samples of urban households; the last two beverages, *pulque* and *aguardiente* or 96% alcohol, are mainly consumed by people in rural areas, or by working-class people in some cities. It is therefore quite understandable that, according to another study, the National Survey of Income and Expenditure of Rural and Urban Households, the two main alcohol beverages reported by inhabitants of poor homes, without inside water and other services, were *pulque* and *aguardiente* or 96% alcohol (44.8% and 52.8% of respondents, respectively) (Medina-Mora, 1999).

The main pattern of alcohol use among males (18%) is of large quantities (5 or more drinks per occasion) on few occasions (once or twice a month); this fact, of people who drink infrequently but to the point of drunkenness, explains the high incidence of acute health and social problems such as accidents and fights related to alcohol, which are quite frequent in Mexico. Using the criteria of *DSM–IV*, it was found that 9% of men and 1% of females between the ages of 18 and 65 years reported signs and symptoms of alcohol dependence. The age group with the highest rate of drinkers is the 30- to 49-year age group.

Although Mexico has a federal law that forbids the sale of alcohol to minors aged 18 or less, it is not properly enforced and alcohol consumption by youngsters is a major cause of concern to parents and authorities. In the latest survey carried out among students 12–19 years old (N = 3,883), alcohol use was reported by a third of the minors interviewed; of these, 5% consumed 5 or more drinks per occasion once a month and 1% did so at least once a week. Three percent said they had been intoxicated during the month previous to the study and 53% reported that they did not know how many drinks would make a person unable to drive a car. Almost 40% drank at home with the consent of their parents, whereas 41% bought alcohol beverages at stores or in discos or bars without being asked for identification (Tapia Conyer, Medina-Mora, & Cravioto, 2001).

METHODOLOGY OF THE STUDY

Various epidemiological and psychosocial studies on alcohol consumption and its impact on health and behavior have been carried out in Mexico. The present study was aimed at observing different types of alcohol use by population groups living in areas where noncommercial beverages were more readily available. Information was obtained from personal diaries with the registration of daily drinking by participants during one month. The study areas were chosen with that criterion in mind. Variables suggested to participating countries were as follows:

- Sociodemographic data: age, gender, marital status, level of education, occupation, and income.
- Drinking practices: type of beverages consumed on each drinking occasion reported in a diary, where respondents drink and with whom, alcohol expenditure, and reasons to drink.
- Drinking effects: intoxication, joy, physical or emotional problems.

Study Sites

The first study site was in the capital of the country, Mexico City, known as *Distrito Federal* or Federal District (DF). The area selected was located in the old part of downtown, a very large and traditional working-class area devoted since pre-Hispanic and colonial times to intensive trade. The area includes old neighborhoods and several large wholesale and retail markets selling food and other goods. The area is also well known for the strong cohesion and solidarity of the community, especially against the authorities, because many illegal activities are carried out by members, such as selling drugs, smuggled or fake CDs and tapes, and clothing and distilled alcohol beverages of well-known brands either illegally imported, stolen, or faked. Although some of the trade takes place inside buildings, the streets are the main trading scene, without payment of rent or taxes; this is part of the landscape of the area. Although police raids are frequent, they have not stopped the illegal activities and it is said that corruption is the name of the game. Many of the merchants live in the area in *vecindades,* very old buildings with a central shared patio and small apartments that are occupied by many families; people have lived there for generations and they operate as part of a network, helping each other in a strong alliance either to avoid the law or to start a new business. Many famous old *cantinas* and *pulquerías* are located in the area, and traditional Mexican food and beverages are sold in the streets (Natera, Tenorio, Figueroa, & Ruíz, 2002). The area is also famous for its culture, expressed in the street graffiti and other means of artistic expression, as well as the use of slang and musical preferences. There are flea markets on the weekends where intellectuals and all sorts of people meet.

Several visits and observation trips were made to the area, together with interviews with the staff of a public treatment center for alcohol dependence located in the neighborhood. This center belongs to the National Institute of Psychiatry. Three *colonias* or big neighborhoods were selected: Centro, Guerrero, and Morelos.

The second study site was located in the state of Hidalgo, in a rural area known as Valle del Mezquital. There, rudimentary roads connect small villages. The area is very poor and lacks many basic services. The inhabitants are mainly *mestizos* or from Indian communities. In this area it was essential to have the support of local health workers. The production and consumption of *pulque* and *aguardiente* are widespread there, and rates of death from liver cirrhosis are among the highest in the country.

Materials

Exercise books of the type used in elementary schools were given to each participant, with their names on the front cover. The first page of each of these "diaries" contained a list of the type of information respondents were asked to include in every entry; there were also explanations on points such as what constitutes a "drinking occasion" to help them to complete their diaries. Field-workers had a register to keep control of the study, containing the name and address of each participant, notes on supervisory visits, the date when the person started to fill the diary, information on specific issues, and other observations.

Sample

Through interviews with staff at the alcohol center in the study area in Mexico

City, contacts were made with local people who were well known and respected in the neighborhood. Among them were two women who had lived in the area for most of their lives and agreed to collaborate in the study. One is a veteran of Alcoholics Anonymous (AA) and very active in the service structure of AA in the area (she makes her living by selling counterfeit DVDs and CDs in the market). The other is a retired merchant. Several meetings were carried out with these contacts to explain the nature of the study and their role in helping the field-workers (a psychologist and a social worker) to select the households and participants. It was very important to visit the homes with the two local contacts to obtain the participation of the community members. Both contact people and field-workers were paid for their work. Several visits were made to households to explain the study to members of each home; visits had to be repeated because nobody was there during the daytime—the children were at school and the adults were out working. It was necessary to visit them at evenings or during the weekends.

Resident members of households at each selected site who were 16 or over and said that they drank alcohol were invited to participate in the filling of the diaries. No requirement for a specific pattern of consumption was imposed. Respondents were asked to make entries in their personal diaries every day, whether they drank or not, during 30 consecutive days. Some people promised to participate but subsequently did not for a variety of reasons: leaving town for a trip, becoming a member of AA, experiencing difficulties in writing, or not being at home on the due date.

Because monetary compensation (the equivalent of US$40) was offered to each participating family, only one member of each household finally entered the sample and completed a diary. Probably, if money had been offered for each diary, more people would have participated. A balanced number of men and women was chosen to complete the diaries. The information provided by the diaries was in many cases very poor, probably because the participants with less education had difficulties writing. In several cases, the field-workers had to visit homes frequently to ensure that people completed their entries for the past days.

RESULTS

Sociodemographic Profile of the Participants

In total, 51 diaries were obtained, 25 in Mexico City and 26 in Hidalgo. In Mexico City, 13 males and 12 females participated, and in Hidalgo, 16 males and 10 females, respectively (see Table 7.1). At both sites more than 60% of participants were aged 44 years or younger. The proportion of young participants (people between 15 and 24 years of age) was higher in Hidalgo (23%) than in Mexico City (12%).

The educational level was somewhat lower among women than men at

TABLE 7.1. Sociodemographic Data

Parameter	Mexico City						Hidalgo					
	Men	%	Women	%	Total	%	Men	%	Women	%	Total	%
Age												
15–24 years old	1	8	2	17	3	12	2	13	4	40	6	23
25–34 years old	3	23	3	25	6	24	8	50	2	20	10	38
35–44 years old	3	23	4	33	7	28	3	19	—	—	3	12
45–54 years old	2	15	1	8	3	12	1	6	1	10	2	8
55–64 years old	4	31	2	17	6	24	2	13	3	30	5	19
Total	13	100	12	100	25	100	16	100	10	100	26	100
Education												
Illiterate	—	—	1	8	1	4	1	6	1	10	2	8
Elementary	4	31	4	33	8	32	2	13	1	10	3	12
Middle/secondary	6	46	5	42	11	44	6	38	4	40	10	38
High school/college	3	23	2	17	5	20	7	43	4	40	11	42
Total	13	100	12	100	25	100	16	00	10	100	26	100
Occupation												
No occupation/retired	2	15	4	33	6	24	4	25	—	—	4	15
Domestic work	—	—	1	8	1	4	—	—	4	40	4	15
Employee	7	54	1	8	8	32	5	31	3	30	8	31
Merchant	3	23	4	33	7	28	1	6	—	—	1	4
Peasant	—	—	—	—	—	—	4	25	—	—	4	15
Professional	1	8	2	17	3	12	1	6	1	10	2	8
Student	—	—	—	—	—	—	1	6	2	20	3	12
Total	13	100	12	100	25	100	16	100	10	100	26	100

both sites, a situation that is also found in the general population. Among participants in Hidalgo there were more men with higher education than those in Mexico. One explanation of that finding may be that in the study area in Mexico City men are mostly involved in merchant activities in the market place. In Hidalgo, on the other hand, if the younger men do not achieve a better level of education their only job option is to remain peasants like their parents, so many of them live there but work at the nearest larger city as employees in businesses or offices.

At both sites a similar proportion of men were married or living with their partners, which was the main status, followed by being single. Among Mexico City women, most were single, followed by those who were married; in Hidalgo, women were equally represented as single and married.

With regard to occupation, men in Mexico City were mostly employees or merchants; in Hidalgo, 25% did not work and the rest were chiefly either employees or peasants. In Mexico City, women who worked were mainly merchants, whereas in Hidalgo they largely worked at home or as employees.

Most of the participants at both sites stated that they owned their homes, although in Hidalgo rented or loaned housing was also reported by a few participants. In terms of their income, two-thirds of the Mexico City participants earned twice the minimum wage, whereas only 24% earned more. In Hidalgo, almost two-thirds had an income higher than twice the minimum wage.

Alcohol Use

At both sites, men reported drinking on more occasions than women (see Table 7.2). In Mexico City, the 13 men participating reported 198 occasions of alcohol use during weekdays and weekends, in similar proportions. In Hidalgo, the 16 men had 142 entries of occasions when drinking took place, but mainly on weekdays. Occasions for drinking among women at both sites were somewhat higher on weekdays than on weekends, but the 12 women in Mexico City reported many more drinking occasions than the 10 women in Hidalgo (147 vs. 58). In Hidalgo, drinking occasions were somewhat more frequent during weekdays, because of *pulque* drinking by peasants when they worked.

Participants reported that they mostly consumed one type of beverage when they drank (83% in Mexico City and 78% in Hidalgo); on remaining occasions they drank more than one type. Compared with the women, a higher proportion of men drank more than one type at both sites.

According to the number of drinking days reported during the 30-day period of completing the diaries, participants at both study sites were allocated to different intervals: 19 participants (37%) reported drinking on 1 to 7 days; 20 participants (39%) drank on 8 to 14 days; and 14 participants (24%) drank alcohol on 15 or more days. This is consistent with one of the drinking characteristics found among the general population—infrequent use but in large quantities per occasion.

TABLE 7.2. Alcohol Use

	Mexico City						Hidalgo					
Parameter	Men	%	Women	%	Total	%	Men	%	Women	%	Total	%
Drinking days												
Weekday	103	52	80	54	183	53	87	61	34	59	121	61
Weekends	95	48	67	46	162	47	55	39	24	41	79	40
Total	198	100	147	100	345	100	142	100	58	100	200	100
Drinking per occasion												
One type of beverage	159	80	127	86	286	83	104	75	52	85	156	78
More than one type	39	20	20	14	59	17	35	25	9	15	44	22
Total	198	100	147	100	345	100	139	100	61	100	200	100

Frequency of use in	Number of days									
both locations	1–7	%	8–14	%	15–21	%	22–30	%	Total	%
Men	11	38	8	28	7	24	3	10	29	100
Women	8	36	12	55	1	5	1	5	22	100
Total	19	37	20	39	8	16	4	8	51	100

TABLE 7.3. Patterns of Alcohol Use by Number of Occasions

	Mexico City						Hidalgo				Total occasions				
Type of beverage	Men	%	Women	%	Total	%	Men	%	Women	%	Men	%	Women	%	Total
Beer	120	52	64	42	184	48	76	44	22	30	196	49	86	38	282
Rum/brandy/ whisky/vodka	48	21	39	26	87	23	15	9	4	5	63	16	43	19	106
Tequila	39	17	27	18	66	17	7	4	14	19	44	11	41	18	87
Sherry/*rompope*	4	2	10	7	14	4	0	0	0	0	4	1	10	4	14
Wine	7	3	3	2	10	3	2	1	0	0	9	3	1	12	
Cocktail	3	1	4	3	7	2	0	0	2	3	3	1	6	3	9
Cognac/*amareto*	3	1	2	1	5	1	0	0	0	0	3	1	2	1	5
Mescal	3	1	0	0	3	1	0	0	0	0	3	1	0	0	3
Pulque	3	1	3	2	6	2	72	42	31	42	75	19	34	15	109
Tepache	1	0	0	0	1	0	0	0	0	0	1	0	0	0	1
Total occasions	231	100	152	100	383	100	172	100	73	100	403	100	225	100	628

As Table 7.3 shows, beer was the beverage consumed on most occasions by men and women in Mexico City (52% and 42%, respectively), followed by distilled spirits such as rum, brandy, whisky or vodka, and *tequila*. Beer was also the main beverage in Hidalgo for men, mentioned in 77 (44%) reports on their drinking occasions, but it was closely followed by *pulque,* which was used on 72 (42%) occasions. Women in Hidalgo reported drinking *pulque* on 31 (42%) occasions, followed by beer on 30% of occasions and *tequila* on 19%. Other beverages, such as wine or cocktails, were mentioned less frequently, and mainly by participants in Mexico City. Among all the participants, men almost doubled the number of drinking occasions reported by women (403 vs. 225). In total, 628 drinking occasions were reported by type of beverage, at both sites and by both men and women during the 30 days of diary recording.

Drinking occasions on weekends (see Table 7.4) represented around 40% of all beer and distilled beverage consumption, and 50% of the few occasions when wine was consumed. On the other hand, *pulque* seemed to be used evenly throughout the week, and only 27% of the drinking occasions were during weekends.

Considering only drinking on weekends by all the sample, analysis of the amount of alcohol consumed by type of beverage revealed that among those drinking beer (115 total occasions reported), the quantity used was up to 1 liter on 41 occasions (36%), on 35 occasions (30%) from 1 to 2 liters, and between 2 and 3 liters on 39 occasions (34%). In the case of distilled beverages (including *tequila* and *mescal*), on most of the drinking occasions the quantities consumed were up to 1 liter on weekends. On 12 of the 29 occasions when *pulque* was consumed (41%) it was taken in large quantities of up to 3 liters.

In Mexico City men recorded 187 diary entries on places where drinking took place and women recorded 135. In Hidalgo there were 127 entries by male participants and only 59 by females. Most of the participants at both sites, both males and females, said that they drank at home. On 62% of the occasions in Mexico City and 44% in Hidalgo men drank at home. Women's

TABLE 7.4. Patterns of Use on Weekends

Main type of beverage	Occasions in both locations		Occasions by quantity in liters						
	Total	%	<0.5	0.6-1	1.1–1.51	1.6–2	2.2–2.5	2.6–3	>3
Beer	115	41	24	17	17	18	8	7	24
Rum/brandy/ whisky/vodka	43	41	27	12	3	1	—	—	—
Tequila/mescal	39	40	25	6	3	1	—	—	—
Pulque	29	27	4	6	3	1	—	3	12
Wine	6	50	3	—	—	2	1	—	—

drinking took place at home on 52% of the occasions in Mexico City and on 56% in Hidalgo. Males drinking in a *cantina* or pub was more common in Mexico City than in Hidalgo (17% vs. 7% of the occasions); women in Mexico City also reported 20% of alcohol use in such public places, against only 5% in Hidalgo. More drinking occasions took place in the workplace in Hidalgo—by both males and females—than in Mexico City. This may be attributed to the drinking of *pulque*, which is traditionally used by peasants in the fields to gain strength and fight thirst. It is common practice for the boss to offer some *pulque* to the workers. In Hidalgo there were some diary reports of drinking on the street, at the market, and in church. At both sites there was also mention of halls for parties as drinking places when celebrations took place.

As to the persons with whom the participants drank alcohol, in Mexico City men made 180 entries and women 136 entries on drinking occasions; in Hidalgo, the corresponding figures were 131 and 56. In Mexico City, most men drank in similar proportions with family or friends; only on 17% of occasions did they drink alone. Women drank mainly with family (52%), followed by friends (33%); on 15% of occasions they drank alone. In Hidalgo, men mostly drank with friends (50%), and then with their families (41%); on just 9% of occasions they consumed alcohol alone. Women in Hidalgo drank almost exclusively with the family (70%), and on remaining occasions with friends; only one occasion of drinking alone was reported.

With regard to reasons for drinking (see Table 7.5), there were differences between the two locations and between men and women at both sites. Men in Mexico City made 80 diary entries on this variable and women 68. For men, celebrations (of weddings or anniversaries, and during civic or religious festivities) were the main motives to drink (43%), followed by soccer games and "because it is a habit" (24%). For women, celebrations were practically the only reason to drink in Mexico City (78%), although there were a few mentions of sadness as a motive (6%). In Hidalgo, there were 79 entries in men's diaries and 42 in women's. Most men reported celebrations as the main reason to drink (49%), followed closely by habit (35%); women reported celebrations and habit as equally important (31% each), and thirst and sadness were mentioned on 14% and 17% of occasions, respectively.

Most participants spent less than one minimum monthly wage for the study period on alcohol, but in Mexico City there were also some people spending more than that. The cost can be considerable considering the low income of most of the participants. On many occasions participants reported that they were invited to drink by friends or relatives. In the case of *pulque* in Hidalgo, some of the participants grew their own *maguey* at home and produced the beverage.

As to drinking effects, in Mexico City 64 men and 74 women made entries on this variable; in Hidalgo, the figures were 79 and 26. More men than women at both sites experienced drunkenness. In Hidalgo, mentions of drunkenness by men were more numerous than in Mexico City (39% vs. 19%),

TABLE 7.5. Reasons for Drinking

	Mexico City						Hidalgo					
Reason for drinking	Men	%	Women	%	Total	%	Men	%	Women	%	Total	%
Celebration	34	43	53	78	87	59	39	49	13	31	52	43
Soccer	19	24	3	4	22	15	8	10	2	5	10	8
Habit	18	23	3	4	21	14	28	35	13	31	41	34
Thirst	—	—	—	—	—	—	4	5	6	14	10	8
Sadness	1	1	4	6	5	3	—	—	7	17	7	6
Traveling	8	10	5	7	13	9	—	—	1	2	1	1
Total	80	100	68	100	148	100	79	100	42	100	121	100

whereas again in Hidalgo more women reported intoxication than in Mexico City (14% vs. 4%). Experience of joy when drinking was recorded in second place equally among males in Mexico City and Hidalgo. More women than men in Mexico City reported feeling joy, but in Hidalgo no women mentioned feeling joyful when they drank. Feeling depressed, aggressive, or with some indisposition was reported by a few participants in both areas, and in a somewhat higher proportion by women in Mexico City.

DISCUSSION

The results of the sample survey confirmed what other studies have shown as to the importance of beer as the main beverage of preference for most people. Beer is manufactured only industrially, and no home-brew has been recorded in Mexico. *Pulque* is still an important drink for peasants in rural areas such as the study site in Hidalgo, and there were even some mentions of its consumption in Mexico City by the working-class participants. For women in Hidalgo, *pulque* was even more important than beer. This surely has to do with the traditional use of *pulque*, its properties, and its availability: It is a drink that is there, in their homes or at their neighbors', it is cheap and nutritious, it calms their thirst, and their ancestors drank it. *Pulque* seems to be consumed mainly by older peasants, whereas the young people now prefer beer and also try to find other jobs or to emigrate.

Drinking was a social and family activity for the study participants. Few people drank alone even if they wanted to get drunk, although in Mexico City drinking alone was reported on some occasions. Perhaps the sense of community can sometimes be lost in the big city. Drinking on the streets is forbidden, and in the city it was not done because the police can arrest drinkers or ask for a bribe. In Hidalgo, however, there was less control and more tolerance toward such practices, even drinking in church, as was reported in some instances there. Of course, men were more prone to drink with other male friends, and women with members of the family.

Although daily alcohol consumption was uncommon, there were several male participants who drank quite frequently during the month of the study, and drinking took place on weekdays as well as during weekends, with many drinking occasions. Among the reasons people gave for drinking alcohol, "celebrations" are similar to the old rituals of the ancient Mexicans in pre-Hispanic times, and even soccer games or bullfights have become a kind of ceremonial gathering for many in this country where drinking is part of almost every encounter. Sadness was mentioned as a motive to drink by some people, mainly women, which points to an interesting aspect of alcohol as a "medicine for the soul." Another reason mentioned by a few participants in Hidalgo was "thirst," a reminder that in places where for centuries potable water was inaccessible, *pulque* was the only drinkable liquid at hand. Even today potable water is very scarce.

The spending on alcohol seems to have been quite high considering the income of the two groups of people. But this also relates to the Mexican culture and traditions and the Roman Catholic character in that even the poorest people will spend everything they have on celebrations and festivities for the saint of their village or the baptism of a child. With regard to drinking effects, drunkenness was quite widespread, and indeed one of the main objectives when people drink in Mexico is to get intoxicated. The idea of drinking to feel relaxed and to enjoy the flavor of a particular brand of beverage is quite outside the general drinking culture: People drink to get drunk. This is a major concern in terms of public health. As reported previously (Medina-Mora, 1998; Tapia Conyer et al., 2001), people cannot (or maybe will not) recognize the difference between moderate and excessive drinking, what is "crossing the line." Feeling joyful was something that women in the rural area of Hidalgo did not experience or report in their diaries; perhaps their life conditions are too harsh, whereas women in Mexico City link drinking with more freedom in their gender status.

There is a rapid change in norms regarding alcohol use—who should drink and how much—mainly in large urban areas. In the past, the drinking behavior of men and women, shaped by a traditional bourgeois culture, reflected the fact that male drinking and intoxication were tolerated and even expected as a sign of manhood, whereas drinking by women was disapproved of. Nowadays, women are drinking more and young girls are becoming intoxicated in discos and bars like their male companions—some say, even more so. The tourist industry and some marketing practices in urban areas are contributing to promote damaging patterns of alcohol use among youngsters, such as drinking too many drinks rapidly in an unsafe setting.

Commentators consider that increases in drinking by urban women in Mexico has to do with their new responsibilities, independence, and roles in society. In rural areas and among the working class, of course, women have always worked, usually more than men, and they drank and got drunk without being censured.

With regard to the purpose of the study concerning noncommercial alcohol, in Mexico some beverages such as *pulque* or *aguardiente* are very much immersed in the culture and traditions of subgroups of the population, mainly those who live in poor conditions or are still not involved with modernity. Such drinks may pose a health risk in themselves, or perhaps the risk may lie in the combination with the other living conditions that affect those who consume them, such as poor diet and lack of services.

As to the potable 96% pure alcohol, it is very toxic but also very cheap, and people mix it with fruit juices or tea; the problem with it is not one of culture or tradition but only poverty and the lack of public policy or law enforcement. In any case, the risks are small when compared with the public health dangers and the actual losses for the public finances of the huge counterfeit market.

Therefore, the issue of noncommercial alcohol should be clearly defined so that all its components can be identified for each society, including cultural, health, industrial, economic, and policy aspects.

REFERENCES

Berruecos, L. (1983). Aspectos antropológicos del alcoholismo en México [Anthropological aspects of alcoholism in México]. In V. Molina, L. Berruecos, & M. Sánchez (Eds.), *El alcoholismo en México* [*Alcoholism in Mexico*] (Vol. 2, pp. 1–16). Mexico, DF: Fundación de Investigaciones Sociales.

Bunzel, R. (1940). The role of alcoholism in two Central American cultures. *Psychiatry, 3*, 361–387.

Corcuera de Mancera, S. (1996). *Entre gula y templanza. Un aspecto de la historia mexicana.* Mexico, DF: Fondo de Cultura Económica.

Instituto Nacional de Estadísticas, Geografía e Informática (INEGI). (2002). *Estadísticas Sociodemográfica.* http://www.inegi.gob.mx

Lozano Armendares, T. (1998). Del chinguirito al ron [From chinguirito to rum]. In *Beber de tierra generosa. Historia de las bebidas alcohólicas en México* [*Drinking of generous land. History of alcoholic drinks in Mexico*] (pp. 128–147). Mexico, DF: Fundación de Investigaciones Sociales.

Medina-Mora, M. E. (1998). Beber en el campo y la ciudad [Drinking in the country and the city]. In *Beber de tierra generosa. Historia de las bebidas alcohólicas en México* [*Drinking of generous land. History of alcoholic drinks in Mexico*] (pp. 207–227). Mexico, DF: Fundación de Investigaciones Sociales.

Medina-Mora, M. E. (1999). *Patrones de consumo de pulque en la zona centro del país* [*Patterns of consumption of pulque in the central zone of the country*]. (Cuadernos FISAC No. 2) (pp. 21–28). México, DF: Fundación de Investigaciones Sociales.

Natera, G., Tenorio, R., Figueroa, E., & Ruíz, G. (2002). Espacio urbano, la vida cotidian y la adicciones. Un estudio etnografico sobre alcoholismo en el centro historico de la ciudad de Mexico. *Salud Mental.* (Vol. 25, No. 4. pp. 12–31.)

Presidency of the Republic. (2002). Speeches and press communications. http://www.presidencia.gob.mx

Rosovsky, H. (1985). Public health aspects of the production, marketing and control of alcohol beverages in Mexico. *Contemporary Drug Problems,* Winter, 659–678.

Rosovsky, H. (2001). Salud pública, disponibilidad y consumo de alcohol: Implicaciones y controversias [Public health, availability and alcohol consumption: Implications and controversies]. In R. Tapia Conyer (Ed.), *Las adicciones: Dimensión, impacto y perspectivas.* [*Addictions: Dimensions, impact and perspectives*] (2nd ed., pp. 169–185). México, DF: Manual Moderno.

Rosovsky, H., & Romero, M. (1996). Prevention issues in a multicultural developing country: The Mexican case. *Substance Use and Misuse, 31*(11&12), 1657–1688.

Sahagun, B. (1969). *Historia General de las cosas de la Nueva España* [*General history of the things of the New Spain*] (Vols. 1–4). México, DF: Porrúa. (Original work published 1582)

SEDESOL. (2002). *La medición de la pobreza en México al año 2000* [*Measurement of poverty in Mexico in the year 2000*]. http://www.sedesol.gob.mx

Soberón, A. (1998). Elixir milenario: El *pulque*. [Millenary elixir: *Pulque*] In *Beber de tierra generosa. Historia de las bebidas alcohólicas en México* [*Drinking of generous land. History of alcoholic drinks in Mexico*] (pp. 29–49). Mexico, DF: Fundación

de Investigaciones Sociales.
Tapia Conyer, R., Medina-Mora, M. E., & Cravioto, P. (2001). Epidemiología del consumo del alcohol [Epidemiology of alcohol consumption)]. In R. Tapia Conyer (Ed.) *Las adicciones: Dimension, impacto y perspectivas* [*Addictions: Dimensions, impact and perspectives*] (2nd ed., pp. 127–138). México, DF: Manual Moderno.
Taylor, W. B. (1979). *Drinking, homicide and rebellion in colonial Mexican villages.* Stanford, CA: Stanford University Press.
UNDP. (2002). *United Nations Human Development Report: Measuring development and influencing policy.* New York: UNDP.
Vargas, L. A. (1999). *El pulque en la cultura de los pueblos indígenas* [*Pulque in the Indian culture*] (Cuadernos FISAC No. 2, pp. 11–20). México, DF: Fundación de Investigaciones Sociales.
Vargas, L. A., Aguilar, P., Esquivel, G., Gispert, M., Gomez, A., Rodríquez, H., Suarez, C., & Wacker, C. (1998). Bebidas de la tradición [Traditional drinks]. In *Beber de tierra generosa. Historia de las bebidas alcohólicas en México* [*Drinking of generous land. History of alcoholic drinks in Mexico*] (pp. 171–202). Mexico, DF: Fundación de Investigaciones Sociales.
Viesca, C. (2001). Bosquejo historico de las adicciones (Historical overview of addictions). In R. Tapia Conyer (Ed.), *Las addicciones: Dimensíon, impacto y perspectivas* [*Addictions: Dimension, impact and perspectives*] (2nd ed., pp. 3–19). Mexico, DF: Manual Moderno.

Chapter 8

Drinking Patterns of Hazardous Drinkers

A Multicenter Study in India

Gaurish Gaunekar, Vikram Patel, K. S. Jacob, Ganpat Vankar, Davinder Mohan, Anil Rane, Surajeen Prasad, Navneet Johari, and Anita Chopra

ALCOHOL USE IN INDIA: THE HISTORICAL CONTEXT

India, although traditionally considered a "dry" or "abstinent" culture, has a long historical association with alcohol use. A review of religious texts, historical accounts, and other manuscripts by Singh and Lal (1979) noted that "there is no cultural tradition in India which could be described as clearly and unequivocally against the use of alcohol in any form and under all circumstances." The earliest recorded history of alcohol use in ancient Indian civilizations reveals that the Dravidians of southern India were familiar with tapping the palm tree for *toddy*, a fermented beverage. The Indus Valley civilization that flourished in northwest India (2000–800 BC) used fermentation and distillation techniques to produce the intoxicating beverages *soma* and *sura*. *Soma* was considered a euphoriant drink for the upper classes and an offering to the Gods. *Sura* was a drink of the warrior and the lower classes, used mainly for relief from physical hardship. Thus, drinking in those periods was limited to certain social classes and occasions and did not find wide acceptance.

The great Indian epics, the Ramayana and the Mahabharata, are replete with references to drinking. Ancient medical treatises by physicians such as Charak and Susruta (dating from 300 AD) make distinctions between normal drinking and excessive drinking, and even describe the beneficial effects of moderate drinking. In general, however, alcohol use, along with consumption of nonvegetarian food, was considered an undesirable characteristic. Despite this, some persons who drank alcohol, such as the warrior castes, were also recognized to be brave and courageous. The *chaturvarna* system divided society into four castes: Brahmins (priests, teachers, and scholars), Kshatriyas (warriors), Vaishyas (peasants and traders), and Shudras (servants, working castes). The norms set by this system placed restrictions on the patterns of alcohol consumption. Total abstention was advised for the Brahmins, whereas occasional drinking was permitted for the Kshatriyas. The traders were permitted to drink for pleasure, whereas Shudras had limited access to alcohol. The upper caste, the Brahmins, permitted other castes to drink alcohol but codified their own prohibitive attitude to alcohol as one way of ensuring that other castes would be unlikely to seek to move up to the class of Brahmins, thus maintaining the Brahmins' exalted status. The origin and spread of Buddhism in India also contributed to the changing and varied societal attitudes to alcohol. Buddhism strictly prohibited alcohol for monks and in the monasteries. The rulers from the warrior class, however, continued to drink alcohol. Thus in ancient Indian society, negative and prohibitive attitudes coexisted with attitudes that even idealized intoxication by alcohol, and alcohol use was one of the criteria for social stratification.

The Moguls, who ruled much of India from the 16th century onward, had a major influence on the drinking culture in India. Although alcohol was often permitted, particularly for the ruling classes, religious edicts from Islam promoted total abstinence from alcohol. These contradictions further accentuated the ambivalence of Indian society about alcohol. The Europeans (French, Portuguese, Dutch, and British colonialists) brought foreign-made liquor to be traded in India. The influence of European colonial rule on the local culture liberalized the consumption of alcohol, especially in the social classes in close contact with the colonizers. The distillation process was introduced followed by the setting up of distilleries. The subsequent introduction of levies and taxes on alcohol meant that alcohol became a source of revenue, which was encouraged by the colonial rulers.

The Indian National Congress, founded in 1885, considered alcohol to be one of the evils of British colonial influence. Leaders such as Mahatma Gandhi favored the total prohibition of alcohol as a part of the independence platform in 1920 adopted at the annual session of the All Indian Congress Committee. Drinking alcohol became a defining feature of what was "bad" in Indian society, under the influence of colonial culture. This stance of the leaders of the freedom struggle was enshrined in the Indian Constitution in 1949 as one of the directive principles of government policy. According to Article 47, "The

State shall regard the raising of the level of nutrition and the standard of living of its people and the improvement of public health as among its primary duties and in particular, shall endeavor to bring about prohibition of the consumption, except for medical purposes, of intoxicating drinks and drugs, which are injurious to health." The implementation of this directive principle has been feeble in most of the states of India. The policy of total prohibition has been followed in the state of Gujarat since 1960 and for shorter periods in some other states such as Haryana, Andhra Pradesh, Nagaland, and Tamil Nadu. However, there is evidence that this policy is rarely enforced consistently, and a flourishing black market in alcohol and the production of noncommercial alcohol are reported in areas where prohibition is enforced (Patel, 1998). More recently, the forces of globalization and the media have had a considerable influence on attitudes toward alcohol consumption.

TYPES OF ALCOHOL BEVERAGES

The production of alcohol occurs in many settings, in both the organized (commercial) and the unorganized (noncommercial) sector. A third category is that of illicit or illegal alcohol. This category is highly variable across India; thus, in some states, all types of alcohol (including those produced by the commercial sector) are illegal, whereas in others, only those produced outside the official alcohol revenue system are illegal. This situation makes any estimate of alcohol production difficult and imprecise. In this study, the category of illicit alcohol is not used in the context of the illegal sale of commercial alcohols. For the purposes of this chapter, the term "illicit alcohol" refers to types of noncommercial alcohol beverages sold outside the official government revenue system and evading quality controls.

Cultural diversity, historical and geographic factors, and variations in regional government policies for alcohol production and use influence the patterns of consumption in the different states in India. Political factors play a key role in influencing the patterns of drinking in these different states. Some states—for example, Goa—have lower excise rates, and the result is the relatively lower cost and liberal use of alcohol beverages there. At the other extreme are states such as Gujarat, which have implemented total prohibition as a policy. How much this has translated into controlled alcohol use is debatable, but it is clear that the policy has boosted illicit alcohol production in this state. In between the liberal attitude to alcohol in Goa and total prohibition in Gujarat is a varied picture of alcohol policies and use in the country. It has been estimated that unrecorded or illicit consumption is at least half of the total recorded consumption. The recorded per-capita consumption is 1.2 liters of absolute alcohol; assuming that this represents only half of actual consumption, the true per-capita consumption is estimated to be close to 2 liters (Saxena, 1999).

Commercial Beverages

Wines are made from a variety of fruits such as peaches, plums, and apricots, but are most commonly produced from grapes. This is a relatively new industry in India. At present, wine is rarely consumed, although there has been an increase in consumption among the rich classes in recent years.

Beer is widely available in India, and there are a large number of breweries. It is available in a variety of strengths (5–9% alcohol content) and is usually sold in 650-ml glass bottles. Recently, smaller bottles and cans have also been marketed. Beer is usually drunk by the middle and upper economic classes.

The word "liquor" is used in this chapter to refer only to distilled beverages. The "Indian-made foreign liquor" (IMFL) category was created for revenue purposes, and consists of Western-style distilled beverages such as whisky, gin, rum, vodka, and brandy. They are sold in bottles ranging in volume up to 750 ml with a maximum permissible alcohol content of 42.8%. Whisky is by far the most popular of the IMFL beverages in India.

Noncommercial Beverages

This is a large group of alcohol beverages, varying from region to region in India. Noncommercial drinks can be broadly classified in two categories: traditional alcohols (fermented or distilled) and illegal alcohols.

Traditional Beverages

Traditional beverages are those that are brewed using local produce. These alcohols may be either fermented or distilled (the latter variety is also sometimes referred to as Indian-made country liquor). In most states, the source of traditional alcohols has been local produce, such as rice, wheat, potatoes, flowers of the Mahua tree, sap from different types of palms, and locally available fruit. The northern and western states of India are sugar-producing areas, and molasses is available cheaply. These are often the substrate for country liquor in these states. Coconut and other palms are common substrates used in south India. An example of a traditional fermented alcohol is *toddy*, which is obtained from the flowers of the coconut or other species of palm. The flowers exude a white liquid with a sweetish taste, which is collected and allowed to ferment. At times yeast is added to hasten the process. The fermented juice has an alcohol content of between 5 and 10%. An example of distilled traditional alcohol is *urrack*, which is produced in coastal states of western India from the fruit of the cashew tree. *Arrack* is a distilled beverage, obtained from paddy or wheat. Jaggery or sugar is added to either of these two cereals and boiled with water. Sometimes sugar cane may also be added. The resulting liquid is allowed to ferment, after which it is distilled.

Distilled traditional alcohols are subject to excise duties, but these are at lower rates than IMFL, and the alcohol produced can only be sold within a specified geographic zone. The licensing system ensures a relative uniformity of the alcohol content, which should not exceed 40%. In states such as Goa, where government policies on alcohol are liberal, the production and sale of noncommercial liquor are subject to few restrictions. The most commonly drunk noncommercial alcohol (*feni*) is distilled by commercial alcohol breweries and is sold in bars and taverns along with IMFL and beer. Thus, strictly speaking, *feni* is commercial liquor but maintains its noncommercial status, thereby evading certain higher taxes.

Illicit Alcohol

This is, by definition, illegal alcohol. However, in states such as Gujarat that have prohibition, all alcohol is illicit. There, as pointed out earlier, we use the term "illicit alcohol" to describe alcohol beverages that evade quality controls. As a result, the alcohol content varies and can be as high as 60%. In order to obtain higher strengths of alcohol, chemicals such as battery acid, urea, and ammonium chloride are added. The reason for addition of additives is unclear, although it is purported that they help increase the intoxication produced by alcohol. In certain regions, by-products from units producing IMFL are used to produce illicit alcohol. The common characteristic of these alcohol beverages is their low cost, which makes them attractive to low-income groups. The lack of regulations makes these beverages health hazards, as evidenced by the regular occurrence of methanol poisoning leading to mortality, blindness, and other morbidity.

THE PREVALENCE AND PATTERNS OF ALCOHOL USE IN INDIA

There has been no national prevalence study of alcohol use. However, there is a considerable body of smaller research studies on alcohol use in a variety of settings (for a review, see Saxena, 1999, or Mohan, Chopa, Ray, & Sethi, 2001). The majority of community-based studies have examined the prevalence of drinking alcohol as an event, rather than the level of alcohol abuse or alcohol-related health or social problems. These surveys have revealed highly variable patterns of use in different regions of the country. The one consistent finding in general population surveys has been the very low rates of alcohol use in females. Alcohol use in males (in the past year) varies from 20 to 75%. Hazardous drinking is a level of alcohol consumption that could prove harmful in the future (Edwards, Arif, & Hodgson, 1981). This concept is important because it describes a population with early alcohol-related problems, which is ideal for delivering brief, preventive interventions. There is little epidemiological research on hazardous drinking in India. A few surveys have examined the prevalence of alcohol dependence; as might be expected, the rates are low, ranging from 0.5 to 3.4% of the population (Saxena, 1999). The data on alcohol production and sale from India show that total absolute alcohol manufactured by liquor companies rose from 169.4 million liters in 1976 to 459 million liters in 1991. These figures do not include the millions of liters of illicit liquor produced, which may vary between 20 and 90% of the official sales depending on the district or area concerned. Data on treatment-seeking populations suggest a remarkable rise in consultations for alcohol-related problems from almost all parts of the country: The number of persons seeking help for alcohol use problems has been recorded to be on the increase since the early 1980s, and more so in the 1990s, in both psychiatric hospitals and the psychiatric services of general hospitals, in the public and private sector. Thus a considerable body of evidence indicates that the level of hazardous drinking in India is likely to be high, although population-based data do not exist. There is even less information on patterns of drinking and, to the best of our knowledge, there are no published studies on the use of noncommercial alcohol in India.

OBJECTIVE OF STUDY

The objective of the study summarized here was to describe drinking patterns in different regions of India, examining both commercial and noncommercial alcohols. The rationale was evident from the literature search, which showed few data examining the use of different types of alcohol. The multicenter design arose from the recognition that considerable variations in drinking patterns were likely to be encountered in different regions of India. The study was designed to be descriptive and to focus on male hazardous drinkers. Men were chosen because they account for the largest proportion of drinkers in the country. Hazardous drinking was chosen because it represented a level of consumption of alcohol that had public health implications.

METHOD

Settings

The four centers chosen were Goa, Ahmedabad, New Delhi, and Vellore.

Goa is in western India; the principal investigator was Dr. Vikram Patel, Sangath Society (Indian coordinating center). The coastal state of Goa was ruled for 450 years by the Portuguese until its liberation in 1961. The locally available cashew fruit and coconut palm sap have traditionally been used to obtain liquor and are the basis for a major part of the noncommercial alcohol consumed in this state. Tourism and migration from neighboring states has led to an increasing demand for alcohol. The state has among the lowest taxes on alcohol in the country and alcohol is freely available from bars and shops throughout the state.

Ahmedabad is in western India, with an urban population; the principal investigator was Professor Ganpat Vankar, B. J. Medical College. Ahmedabad is the main commercial city of the western state of Gujarat. Gujarat is the only state in India with an alcohol prohibition policy at the time of writing of this chapter (Patel, 1998), which effectively prevents legal distribution and consumption of alcohol for most people. Thus, virtually all alcohol available is illicit in the true sense. Noncommercial production of liquor is typically in small shacks hidden from the eyes of the authorities. The liquor is made from refined sugars, with some production from palm sap. A thriving black market for the sale of IMFL also operates.

New Delhi is in northern India, with an urban population; the principal investigator was Professor Davinder Mohan, All India Institute of Medical Sciences. New Delhi is one of the world's largest cities and is the country's capital city. Alcohol beverages are freely available, with certain restrictions on distribution. Because of the urban setting, natural resources for alcohol production are meager. The commonest variety of noncommercial alcohol is illicit liquor obtained as a by-product of the IMFL distilleries.

Vellore is in southern India, with a rural population; the principal investigator was Professor K. S. Jacob, Christian Medical College. Vellore is a large town in the southern state of Tamil Nadu. Although alcohol policies are more relaxed compared to states such as Gujarat, the availability of commercial spirits is restricted. For instance, IMFL is freely available in Vellore, but shops selling such commercial liquor are not found in the rural areas except in some villages near the state highway. The noncommercial alcohol produced is made from refined sugars with additives such as tree barks and various chemicals. There are a few villages in the vicinity of Vellore whose economy is said to depend on illicit brewing. Illicit alcohol is easily available in the villages of the Kaniyambadi block where the study was conducted, which is close to Vellore town.

Study Design

The study was in two stages.

The first stage was a qualitative study aimed at generating information about noncommercial alcohols in each study setting. The research team visited the areas where the study was to be conducted. Key informants were contacted, rapport was established, and focus-group discussions (FGDs) were conducted. The focus-group participants included members of the population being sampled, families of drinkers, and key informants in the locality. Two to three FGDs were conducted at each center. The objective was to obtain information regarding the common names of the local alcohols, their contents, sources, availability, and patterns of use.

The second stage was a descriptive study aimed at describing the patterns of alcohol consumption in hazardous drinkers. Individual subjects who were known to be heavy drinkers, on the basis of either clinical assessment or earlier structured interview assessments, were purposively sampled to participate in the study. Subjects who consented were interviewed with the Alcohol Use Disorders Identification Test (AUDIT), a 10-item screening questionnaire developed by the World Health Organization (WHO) for detection of hazardous drinking (Babor, de la Fuente, Saunders, & Grant, 1992). It has been validated and used in cross-national studies in a number of countries, including India. WHO prescribes a cutoff score of 8 on the AUDIT for higher sensitivity in detecting hazardous drinking. Subjects who scored 8 or more were recruited for the study.

The sample selected for this study consisted of about 50 male hazardous drinkers at each of the four centers who were purposively invited to participate in the study. At the Goa center, the sample consisted of male industrial workers in a shipbuilding company. At the New Delhi center, the sample consisted of traditional performers residing in the city slums. At the Ahmedabad center, the sample comprised members of a nomadic tribe, laborers, and unskilled workers. At the Vellore center, the sample was from a village where the inhabitants were mainly stone cutters and laborers. In each study setting, the research team had previous contact with the study population. For example, in Goa the sample was drawn from a population of 984 industrial workers among whom the same research team had conducted an earlier study. At the other centers, the research teams were either part of the general health services being provided to the populations concerned (e.g., Vellore) or had been providing alcohol treatment services in the area (e.g., Ahmedabad).

Two types of data were collected from each subject.

Structured Sociodemographic and Alcohol History Interview

Subjects were interviewed regarding sociodemographic details such as age, marital status, religion, occupation, and income. Information on drinking history included family history of drinking, age of starting drinking, common brands consumed and preferred brands, concomitant tobacco use, attempts at abstinence or cutting down, effects of drinking on work, and family interpersonal relations.

Drink Diary

In this study, a drink diary is a 1-day record of drinking behavior on a specific day. After the initial interview the research workers met the subjects at weekly intervals. Information about drinking on each day in the past week was collected in drink diaries. Thus, on average, records for 28 days were collected for each subject during four visits. However, at one center (Vellore), records were maintained for up to 6 weeks. The first item in the drink diary was whether alcohol had been consumed on that day. If yes, the following questions were asked:

- Day of the week?
- Where?
- With whom?
- Time of starting and ending the drinking sessions and the total time spent drinking?
- Types and amount of alcohol consumed (information on local alcohols was region specific and derived from the earlier focus-group discussions)?
- Amount of money spent?

Analysis

The data from all four study centers were pooled at the coordinating center in Goa. The data were screened for discrepancies and the respective centers contacted for clarifications when required. Drink diary data were analyzed by days. The data were entered using the Epi6 program, a software program of the WHO. The data was analyzed using SPSS for Windows, a statistical program.

RESULTS

Stage 1

The focus-group findings are summarized in Table 8.1.

Stage 2

Sociodemographic Characteristics

In total, 192 subjects were recruited from four centers. The sociodemographic features of the sample (Table 8.2) show that the subjects at all four centers were very similar in terms of their age. Subjects at three centers were similar in terms of household size, whereas the New Delhi sample had relatively more members in the household. The Goa sample consisted of industrial workers, and this was reflected in their higher education and income as compared to the other centers. The monthly income of the subjects was uniform at the Goa center, which can be explained by the fact that all subjects were workers in the same industry. The total family income at all centers was close to the subject income, probably because males are the main breadwinners in most parts of India. The proportion of married men was higher at the Goa, Vellore, and New Delhi centers than at Ahmedabad. Hindus formed by far the largest proportion

TABLE 8.1. Characteristics of Focus Groups Based on Discussions at Four Centers

Characteristics	Goa	Vellore	Ahemdabad	New Delhi
Number of groups	2	2	2	3
Categories of participants	(1) Industrial workers (2) Hospital attenders	(1) Manual laborers and stone cutters (2) Extension health workers	(1) Inhabitants of a local slum area (2) Members of a nomadic tribe	(1) Alcohol users in a colony of traditional performers (2) Wives of alcoholics (3) Community leaders
Availability of alcohol	Commercial and noncommercial alcohol freely available	Commerical and non-commerical alcohol freely available	Illicit liquor due to alcohol prohibition	Commercial and non-mercial alcohol freely available
Names of non-commercial alcohol beverages	(1) Cashew *feni* and its modifications, e.g., *doosri, guchiri, aale, velchi* (2) Coconut *feni*, also called *maad* (3) *Sur*, fermented coconut sap (4) *Urrack*, the first distillate of cashew juice	*Sarayam, naatu sarayam, pattai sarayam, pattai, thani, salpatt, sarak sarayam, naatu*	*Thaili, tharra, desi, daru*	*Deshi* (generic name for country liquor), New Delhi No. 1, *mastana, aspara, seera, shaukeen*, etc.
Packaging	Sold in measures of pegsor by the bottle and its fractions, similar to commercial alcohol	Quantities of about 175 ml, sold in glasses	Sold in polyethylene bags of 200 ml capacity	Plastic pouches of 200 ml
Components	Additives such as ammonium chloride and sugar are added to increase alcohol production; herbs and spices are added to cashew *feni* to produce products such as *doosri, guchiri, aale*, and *velchi;* adulterants rarely used include urea and battery acid	Jaggery, fermented grapes and bananas, bark of trees; adulterants such as battery acid, ammonia, urea, sulfates, spirit, and matchsticks are also used		

of the sample; the ratio of Hindus to non-Hindus was lowest in Goa, where Christians make up a substantial proportion of the population. In Vellore, most subjects came from joint or extended families, whereas the majority of subjects at the remaining centers lived in nuclear families. Hunger, an indicator of severe economic difficulties, was more frequent in the Vellore and New Delhi samples, which is consistent with the lower mean incomes for these centers.

Drinking History

Table 8.3 shows considerable variations in the drinking histories of the subjects at the four centers. AUDIT scores were highest in the New Delhi sample and lowest in the Goa sample. This difference across the four centers was reflected in the reported rates of events associated with alcohol dependence as

TABLE 8.2 Sociodemographic Characteristics

Variable	Goa (n = 55)	Vellore (n = 41)	Abmedabad (n = 50)	New Delhi (n = 46)	All centers (n = 192)
Age (years)					
Mean	43.8	41.7	37.1	38.1	40.3
S.D.	6.8	8.8	9.8	13.2	10.2
Education (years)					
Mean	8.6	3.4	6.8	0.98	5.2
S.D.	3.8	3.3	3.8	2.2	4.5
Number of family members					
Mean	4.4	4.6	4.2	5.9	4.7
S.D.	1.3	1.5	2.4	3.4	2.4
Total monthly family income (Indian rupees)					
Mean	7647.2	2348.7	4774.1	3502.4	4804.7
S.D.	1858.8	978.1	3506.7	2136.4	3025.0
Subject monthly income (Indian rupees)					
Mean	7132.7	1848.7	3886.6	2883.7	4203.8
S.D.	1659.4	822.7	2137.4	1550.1	2658.7
Gone hungry in the past month (%)	4.0	22.0	14.0	24.0	15.0
Marital status					
Married (%)	98.0	98.0	78.0	91.0	91.0
Religion					
Hindu (%)	56.0	93.0	92.0	89.0	81.0
Family type					
Joint (%)	18.0	90.0	42.0	35.0	44.0

Note. Approximate exchange rate is US$1 = Rs 48.

TABLE 8.3. Drinking Histories of Study Participants

Variable	Goa (n = 55)	Vellore (n = 41)	Abmedabad (n = 50)	New Delhi (n = 46)	All centers (n = 192)
AUDIT scores					
Mean	11.1	15.4	16.2	22.9	16.2
95% confidence interval	10.3–12.04	13.4–17.4	14.8–17.6	20.9–24.9	15.1–17.1
Age at first drink					
Mean	21.6	23.3	21.5	18.6	21.3
S.D.	5.9	9.1	7.9	8.1	7.8
Age when beginning regular drinking (years)					
Mean	28.4	29.2	25.3	21.6	26.2
S.D.	6.8	8.5	7.3	7.9	8.1
Maximum abstinence in past year (days)					
Mean	48.4	41.4	82.8	54.6	58.2
S.D.	68.6	31.8	94.3	74.7	75.5
Introduced to drinking by peers (%)	76.0	44.5	67.0	13.0	51.0
Ever tried to cut down (%)	80.0	29.0	58.0	72.0	62.0
Most frequent drink					
Commercial	67.0	15.0	10.0	9.0	28.5
Noncommercial	33.0	85.0	90.0	91.0	71.5
Most preferred drink (%)					
Commercial	65.5	28.0	12.0	14.0	33.0
Noncommercial	34.5	72.0	88.0	86.0	67.0
Morning drinking (%)	5.5	56.0	12.0	70.0	33.0
Binge drinking (%)	13.0	56.0	18.0	56.0	33.0
Help seeking (ever) (%)	4.0	17.0	8.0	13.0	10.0
Smokes cigarettes (%)	13.0	76.0	54.0	74.0	69.0
Chews tobacco (%)	11.0	41.5	60.0	6.5	38.5
Drinker in the family (%)	34.5	12.0	28.0	76.0	30.0
Problems with police in past 3 years (%)	5.5	12.0	16.0	6.5	10.0
Fracture in past 3 years (%)	11.0	46.0	14.0	13.0	20.0
Hit wife/family member in past year (%)	7.0	39.0	24.0	52.0	29.0
Fought at work in past year (%)	16.0	10.0	36.0	20.0	21.0

well; thus, early-morning drinking and binge drinking were least common in the Goa sample and most common in the New Delhi sample. The onset of drinking behavior was in the early twenties in three centers and in the late teens at the New Delhi center. Subjects were most commonly introduced to drinking by peers at three centers; about a third of all subjects, and up to 67% of subjects in New Delhi, reported that they had started drinking on their own. Although commercial drinks were preferred and most commonly consumed in Goa, noncommercial alcohol was the most preferred and commonly used type of alcohol at other centers. New Delhi subjects had the highest proportion of drinkers in the family, followed by Goans; the most commonly cited drinker in the family was a spouse for the Vellore subjects (57%), whereas a sibling or a parent was cited in Goa (84%), New Delhi (49%), and Ahmedabad (43%), respectively. A significant proportion of subjects had experienced an adverse impact on their social and physical health. Thus rates of tobacco use and experience of violence, injury, and trouble with the police were high at all centers, especially in New Delhi and Vellore.

Drink Diary

Data were obtained for a total of 5,390 days from the four centers (Goa 1,307; Vellore 1,589; Ahmedabad 1,456; New Delhi 1,038). Drink diary findings have been summarized in Table 8.4 and Figures 8.1–8.3. The key findings are that the proportion of days on which drinking occurred (drink days) differed at the four centers; Ahmedabad, Vellore, and New Delhi had much higher proportions than Goa. These differences probably reflect the fact that the samples at the three centers other than Goa had significantly more severe levels of hazardous drinking, as suggested by the higher AUDIT scores. IMFL and beer were the most commonly drunk alcohol in Goa, while noncommercial alcohols were the most commonly consumed alcohols at the other centers. The total units consumed were highest in New Delhi, followed by Ahmedabad, again in line with the AUDIT scores, which were highest for these two groups. Drink days were relatively evenly spread throughout the week, with a peak during the weekend (Figure 8.1). Data for New Delhi are not included here because of inconsistent reporting of drinking on Sundays, because many subjects would leave New Delhi to perform in other towns on Sundays. The time of starting to drink was most often in the evening at all centers. However, a quarter of the drink days at the Goa and Ahmedabad centers started in the morning. In New Delhi, early-morning drinking was seen on about 35% of drinking days (Figure 8.2). Almost all the Ahmedabad subjects and a major proportion in Goa drank in bars; in Vellore drinking was done in public places, which were common meeting points for the community; in New Delhi drinking at home was found to be most common. Most of the drinking was done alone in Goa, New Delhi, and Vellore. In Ahmedabad, half of the subjects drank alone and the other half with friends or relatives. The total units of alcohol consumed on drink days varied between the centers. The amounts of noncommercial alcohol consumed were much higher in terms of alcohol units than amounts of commercial alcohol. The amount of money spent in rupees on the drinking days was higher in New Delhi, Vellore, and Ahmedabad than in Goa.

TABLE 8.4. Drinking Patterns From the Drink Diaries

Variable	Goa	Vellore	Ahmedabad	New Delhi	All centers
Proportion of drinking days (%)	49.0	82.0	68.0	72.0	68.0
Drinking location (%)					
Home	37.3	8.5	2.5	73.0	25.0
Bar	59.3	3.0	95.0	10.0	39.0
Friend's place	2.3	1.0	0.5	8.5	3.0
Workplace	0.8	7.0	2.0	5.0	4.0
Public place	0.2	80.5	0.0	3.5	29.0
With whom drank (%)					
Alone	81.5	86.0	49.0	73.0	72.5
With friends/relatives	17.1	11.0	45.5	23.0	24.0
Both	1.4	3.0	5.5	4.0	3.5
Proportion of drinking days when beer consumed (%)	16.1	0.5	6.6	9.6	6.4
Proportion of drinking days when IMFL consumed (%)	53.2	7.3	0.3	3.9	12.8
Proportion of drinking days when noncommercial alcohol consumed (%)	34.7	92.5	92.6	89.2	82.1
Units of beer consumed					
Mean	3.50	1.30	6.70	4.10	4.10
S.D.	1.06	0.58	3.58	1.73	1.65
Units of IMFL consumed					
Mean	4.20	6.30	6.00	11.80	5.10
S.D.	1.74	3.01	0.00	8.17	3.38
Units of noncommercial alcohol consumed					
Mean	5.40	5.60	10.50	14.50	8.90
S.D.	3.39	2.29	9.83	10.77	7.71
Total units consumed on the drinking days					
Mean	4.78	5.70	10.60	13.90	8.32
S.D.	2.50	2.38	11.38	10.67	7.62
Rupees spent on drinking days					
Mean	27.40	36.30	35.09	35.20	34.04
S.D.	16.55	19.62	34.28	29.02	29.96

Note. The number of units is calculated using the total number of days on which drinking occurred as the denominator.

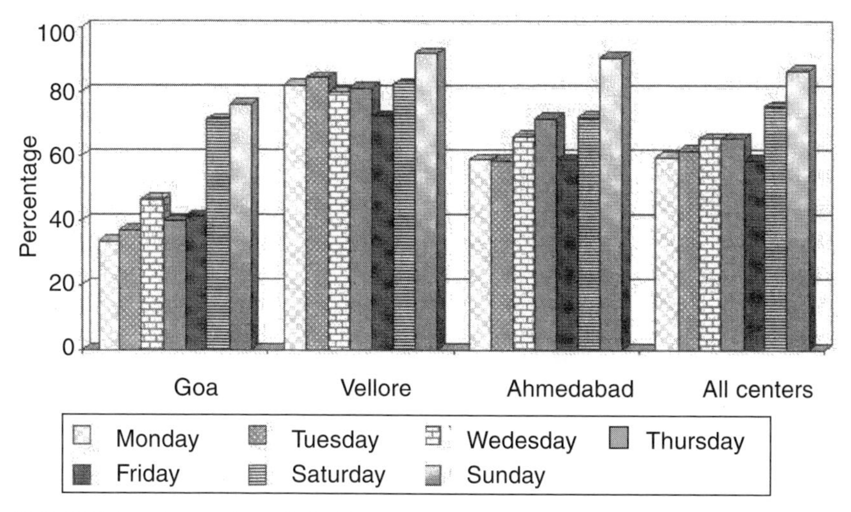

FIGURE 8.1. Proportion of drinking days by day of the week for three centers.

DISCUSSION

This study was a multicenter study of patterns of drinking in a specific category of the population: male hazardous drinkers as defined by the AUDIT questionnaire in India. The study was done in two stages. The first stage consisted of focus-group discussions in the four study areas to determine the various terms in common use to describe noncommercial alcohol and the manner

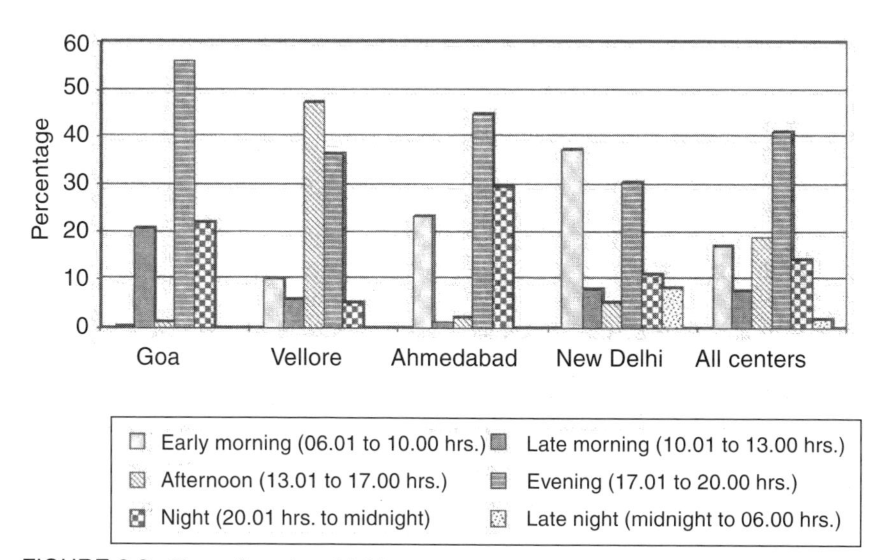

FIGURE 8.2. Time of starting drinking.

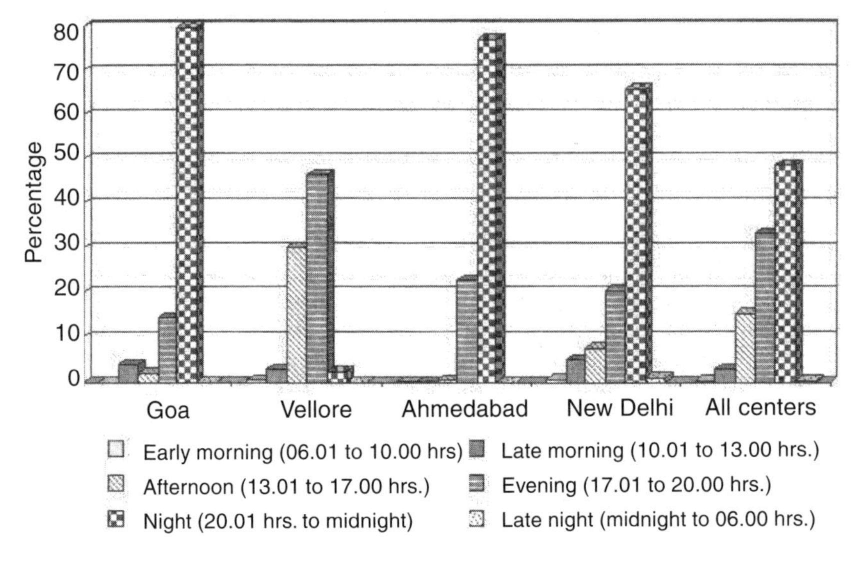

FIGURE 8.3. Time of stopping drinking.

in which they were sold. The second stage was a descriptive study of the drinking patterns in hazardous drinkers at each center, carried out by keeping records for days in 4 to 5 consecutive weeks for each subject using a daily drink diary. Information on the sociodemographic profile and drinking history of each subject was collected through an initial interview with the subject.

The key findings in the focus-group discussions were that, except for the Ahmedabad center, both commercial alcohol and noncommercial alcohol were freely available, with a few restrictions as in New Delhi and Vellore or no restrictions as in Goa. In Ahmedabad, because of prohibition, alcohol drinks of all varieties, including commercial alcohol beverages, were regarded as illegal. Despite this, participants in the discussions reported a wide variety of noncommercial alcohol beverages. The types varied from center to center and had their own unique nomenclature. Commercial alcohols, in contrast, were fairly uniform in their availability at all centers. The measures in which the noncommercial alcohol was sold varied: Goa had the conventional peg and bottle measures, whereas in the other centers alcohol was sold mostly in plastic pouch or glass measures. Noncommercial alcohol was distributed in Goa through bars, taverns, and liquor shops as well as privately in homes. At the other centers, distribution was through shacks, which are small unofficial distribution points, that sold these beverages exclusively. It was commonplace to use adulterants and additives to enhance the alcohol content of the liquor as well as to increase the quantity. Thus, noncommercial alcohol was mostly locally brewed, with distribution limited to the locality; the alcohol content, raw materials used, and names used to describe the liquors varied from center to center. The noncommercial alcohol at two out of four centers (Vellore and Ahmedabad) was illegal, whereas in Goa it was legal, and in New Delhi both legal and illegal varieties of such alcohol were available.

The collection and evaluation of drink diary data were completed at all centers. The sociodemographic characteristics of the samples reflect the considerable variations in these indicators among the different states of the country. Goa, for example, has one of the highest per-capita income levels and literacy rates in India. This, along with the fact that the sample in Goa consisted of industrial workers, accounted for the finding that the Goa sample had a higher educational status, higher income levels, and a lower incidence of hunger due to lack of money. Cultural differences among centers were also evident in the fact that although Hinduism was the most common religion in New Delhi, Vellore, and Ahmedabad, Christians formed a significant proportion of the study sample in Goa. Most of the subjects were married. The samples did not differ much as to age structure and household size. The Vellore study setting was rural, which explains the higher proportion of joint families, the lower education level, and lower family incomes, as well as the higher rates of hunger in this sample. The New Delhi center reported a high proportion of subjects with lower educational status, lower family incomes, and higher rates of hunger; this is probably because the study was conducted in city slums with traditional performers as their main population.

Most subjects at all centers started drinking early in their late teens to early twenties. The regular drinking age was usually the mid twenties to early thirties. An interesting finding was that in Goa, with its liberal drinking culture, peers introduced most subjects to alcohol, whereas relatives were key figures for subjects in New Delhi and Vellore. More importantly, a substantial proportion of subjects at all centers claimed that they had not been introduced to drinking by anyone specific. More than half of all subjects had tried to abstain or cut down their consumption. In Goa, where commercial alcohol is freely available and taxes on alcohol are lower, IMFL and beer were preferred to noncommercial alcohol; in New Delhi, Vellore, and Ahmedabad, in contrast, noncommercial beverages were preferred and most frequently consumed. This can be explained by the fact that IMFL is barely available in Ahmedabad, and in Vellore and New Delhi it is not as freely available as in Goa. The difference in cost between commercial and noncommercial alcohol at these centers is also quite high, because the state government taxes are high. A family history of drinking was most frequent among the New Delhi subjects and least so in Ahmedabad and Vellore. Whenever a subject had such a family history, the drinker was most often a sibling in Goa and New Delhi or a parent in Ahmedabad. Other than in Vellore, no sample identified a spouse as the drinker. The relatively high reported rate of drinking by spouses is probably unique to the population sampled in Vellore, where alcohol consumption and dependence have been reported among women laborers working in the local stone quarries. This is particular to this subgroup of people and cannot be generalized to the population as a whole, where mainly men consume alcohol. Tobacco use (smoking and chewing) was common in Vellore and Ahmedabad, whereas only smoking was common in New Delhi. Help-seeking behavior was least common in Goa, and comparatively higher rates were noted at other centers. This was surprising, since the Goa sample was recruited in an industrial setting with on-site medical facilities. All indicators used in this study for the impact of drinking on the social and family lives of the subjects suggested serious impairment of social and occupational health at all centers.

A key finding from the drink diary data was the high number of units of alcohol consumed on drink days especially in Ahmedabad and New Delhi, which correlates with the higher AUDIT scores reported for these subjects. IMFL and beer were the most commonly consumed alcohols in Goa, whereas noncommercial alcohol was the most commonly consumed at other centers. An association among higher AUDIT scores, higher proportion of noncommercial alcohol use, higher consumption in terms of units of alcohol, and more frequent reporting of alcohol dependence features may be inferred. A majority of the subjects started drinking in the evening and stopped by nighttime (20:00 to midnight). Drinking in bars was by far the most common in Goa and Ahmedabad; in Vellore people commonly drank in public places, which were probably common meeting points for the community of stone cutters and laborers who were the study subjects. The tendencies toward higher rates of drinking at home and of early-morning and late-night drinking were unique to the New Delhi subjects. Drinking alone was common at all centers. This is a risky drinking pattern, because drinking is thus primarily aimed at intoxication, with social reasons for drinking being almost nonexistent. It was notable that the amounts of noncommercial alcohol consumed in terms of alcohol units were higher than those of commercial alcohol, even in Goa where these beverages were not the most frequently consumed form of alcohol. The amount of money spent in rupees on the drinking days was lowest in Goa, in contrast to the income levels, which were highest; this finding is a possible reason why "going hungry due to lack of money" was most often seen at the other centers.

A detailed search of the literature did not reveal any prior research in India on patterns in hazardous drinkers and/or studies focusing on noncommercial alcohol in particular. The comparison of the findings of this study with existing information is thus difficult. A key finding of the study was the contrast between Goa and the other three centers in terms of preferred consumption of commercial alcohols. Goa is a state where the distribution channels for IMFL and beer do not differ greatly from those of noncommercial alcohol. Also, because excise duties for IMFL are not high, the cost difference between IMFL and noncommercial alcohol is much lower than in other states. Cashew *feni*, which is the noncommercial alcohol with the highest consumption in Goa, is produced and sold on a commercial basis today. Thus drinkers have an unrestricted choice between noncommercial alcohol and the preferred, commercial liquor. Noncommercial alcohol consumers are mainly those who are traditional *feni* drinkers or those who drift to the use of noncommercial beverages because of the increasing quantities they consume and the associated financial constraints. It is noteworthy that there have been hardly any reports of poisoning by noncommercial alcohol in Goa, which suggests that most of these beverages are produced with good quality controls. In contrast, the other three centers have a thriving market for noncommercial alcohol, much of which is of the illegal variety. Vellore and New Delhi have less liberal policies on the availability of alcohol and IMFL is much more expensive, leading to greater use of noncommercial beverages in these centers. Ahmedabad is the commercial capital of a state with a prohibition policy; thus, IMFL is not legally sold at all.

This study has given us new insights into the patterns of alcohol consumption in male hazardous drinkers in diverse centers in India. The main finding is that there is wide variation in the drinking patterns at the four centers, with the most notable differences being related to the proportion of use of noncommercial and commercial alcohols. State policies on the production, quality control, taxation, and distribution of alcohols are likely to be the main reason for these variations.

Of those policies, perhaps the most controversial is that of prohibition. In the state in which it has been enforced (Gujarat), prohibition has resulted in restrictions on the sale of any alcohol, but it has also been associated with the illegal production and consumption of noncommercial alcohol. The cheaper noncommercial alcohol beverages are produced from unhealthy sources through unhealthy processes; they have an unrestricted alcohol content and contain other contaminants, which have serious health implications. This is evident from the regular incidents of poisonings and morbidity among their users. The criminalization of alcohol production and sale allegedly has support in the political system and the police forces in some states. A more liberal policy is required for alcohol use, alongside strict enforcement of standards for the production and distribution of noncommercial alcohol. Low or absent excise duties for noncommercial alcohol translate into low prices for such beverages and lead to higher quantities being consumed. Thus, raising the rates of excise duty would make them less attractive to the low-income groups who are their major consumers, reducing their consumption and, in turn, rates of hazardous drinking in the population. Any legislative action must be combined with programs aimed at educating people about the hazards of alcohol and teaching those who choose to drink the appropriate modes of consumption, including moderate drinking levels. The media, which are often active in health education in fields such as family planning and HIV/AIDS, should be encouraged to cover alcohol-related problems as well. Because a finding of our study was that a majority of our subjects in some of the centers tended to drink in bars, these could be a possible venue for educational activities, so that they reach those who need them the most.

In conclusion, drinking patterns in hazardous drinkers in some of the study centers in India were significantly associated with noncommercial alcohol consumption. Variations in the preferences for particular types of alcohol between centers were possibly related to the socioeconomic levels of the subjects and to the costs and ease of access of different types of alcohol in the study areas. Thus, the lower costs, relatively cheaper commercial alcohols, and higher socioeconomic background of subjects at the Goa center may be the primary reasons why these were the preferred alcohol beverages, in contrast to the subjects at other centers. Public health policies aimed at reducing alcohol-associated morbidity must cover both commercial and noncommercial alcohol. Although our study was unable to define the association between dependence and use of cheaper noncommercial drinks, the findings of higher rates of noncommercial alcohol drinking and signs of dependence at two centers suggest a pattern that needs further investigation. Only population-based studies can describe the scale of the problem of hazardous drinking and associated morbidities and the relative contribution and impact of different varieties of beverage alcohol. The descriptive study that has formed the basis for this chapter suggests the need for such population-based research.

REFERENCES

Babor, T. F., de la Fuente, J. R., Saunders, J., & Grant, M. (1992). *AUDIT: The Alcohol Use Disorders Identification Test: Guidelines for use in primary healthcare*. Geneva: World Health Organization.

Edwards, G., Arif, A., & Hodgson, R. (1981). Nomenclature and classification of drug and alcohol-related problems—A World Health Organization Memorandum. *Bulletin of the World Health Organization, 59*, 225–242.

Mohan, D., Chopra, A., Ray, R., & Sethi, H. (2001). Alcohol consumption in India: A cross-sectional study. In A. Demers, R. Room, & C. Bourgault (Eds.), *Surveys of drinking patterns and problems in seven developing countries* (pp. 103–114). Geneva: World Health Organization.

Patel, V. (1998). The politics of alcoholism in India. *British Medical Journal, 316*, 1394–1395.

Saxena, S. (1999). Country profile on alcohol in India. In L. Riley & M. Marshall (Eds.), *Alcohol and public health in 8 developing countries* (pp. 37–60). Geneva: World Health Organization.

Singh, G., & Lal, B. (1979). Alcohol in India. *Indian Journal of Psychiatry, 21*, 39–45.

Chapter 9

Moonshine: Anthropological Perspectives

Linda A. Bennett

An anthropological perspective on the cross-national six-country study of the patterns and effects of noncommercial alcohol consumption is valuable for a number of reasons. To begin with, anthropologists approach research questions and application issues from a broad vantage point somewhat distinct from other disciplines. Second, they encourage a more inclusive interpretation of data, often in novel terms. And third, an anthropological perspective can lead to innovative ideas for future research.

The term "noncommercially produced alcohol beverages" is a broad rubric encompassing drinks typically referred to as "moonshine" or "illicit" or "illegally" produced substances. For simplicity's sake, such beverages are referred to as "illicit" in this chapter. Studying illicit beverages in different cultures focuses on the demand for certain drinks on the part of the general populace and on whether or not authorities approve of their production, distribution, and consumption. Once the more popular illicit beverages are identified for a given culture, questions about who consumes these particular substances most often and under what circumstances can be addressed.

DEFINING AN ANTHROPOLOGICAL APPROACH

What is an anthropological perspective? Anthropologists take a particularly holistic viewpoint in their approach to conducting and interpreting research results. Thus, human behavior, beliefs and symbols, language, social relationships, and physical artifacts and their meaning are encompassed within an anthropological perspective. This has traditionally been the case in anthropological studies of alcohol beverage consumption and related problems.

There is a convention in anthropology generally, and in alcohol studies specifically, to question basic assumptions and to challenge the contemporary thinking about a research or clinical problem. Thus, it is not unusual for debates to arise within anthropology and between anthropologists and colleagues in other disciplines about alcohol consumption patterns and their impact. Anthropologists have a strong tendency to stress cultural variation within and across cultures in the behavior and meaning of alcohol beverage consumption and alcohol-related problems (e.g., Heath, 2000). Rather than generalizing across cultures, anthropologists are inclined to emphasize differences between cultures and subcultures. Furthermore, they place a particular importance on cultural context in advancing ideas to explain differences in patterns of consumption and alcohol-related problems.

Methodologically, anthropologists have drawn on qualitative or ethnographic means for conducting research more often than on quantitative methods, but there has been a shift in recent decades toward combining qualitative and quantitative methods (e.g., Ames, Delaney, & Janes, 1992; Marshall, Ames, & Bennett, 2001; Strunin, 2001). Anthropologists firmly hold that they should have firsthand contact with thc individuals and groups under study and contend that this is a critical aspect of collecting valid data. This position is consistent with a long-standing and significant emphasis on original fieldwork in anthropology and on the value of collecting and interpreting new empirical data in the light of existing data.

Areas of Debate in Anthropology

How does this broad holistic perspective apply to the study of alcohol consumption and alcohol-related problems? Certain areas of debate about patterns and variation within and across cultures stand out: (1) a problem versus "normal drinking" focus in the research and writings of anthropologists (e.g., Hill, 1985; Hunt & Barker, 2001; Room, 1984); (2) a biological versus cultural explanation for the etiology or expression of problem drinking (e.g., Leland, 1976; Levy & Kunitz, 1974; also Kunitz & Levy, 1994, regarding American Indian drinking); (3) relatively greater emphasis on an "emic," or insider's, angle for understanding drinking and drinking-related problems, rather than entirely from the "etic" or outsider's position; (4) cultural influences on drunken comportment (MacAndrew & Edgerton, 1969; Room, 2001); and (5) use of cross-cultural comparisons to demonstrate the broad variability in patterns of drinking and drinking-related problems.

Cross-Cultural Considerations

Anthropologists have been major contributors to cross-cultural descriptions and analyses of alcohol beverage consumption and consequences for decades. Such contributions have taken different forms. In some instances, anthropologists have edited already-published materials regarding drinking in different cultures and provided an overview of the cross-cultural patterns (e.g., Marshall, 1979), or they have solicited papers and chapters around a given theme (e.g., Bennett & Ames, 1985; Gefou-Madianou, 1992; Heath, 1995; Marshall, 1982; Marshall et al., 2001; Riley & Marshall, 1999). In other cases, anthropologists draw from published material to write their own cross-cultural comparisons around certain themes or theoretical ideas (e.g., Heath, 2000; MacAndrew & Edgerton, 1969). It is a misnomer to refer to these works as solely "anthropological" because virtually all are done in collaboration with colleagues from a variety of disciplines. Arguably some of the most useful cross-cultural analyses have come out of planned studies in different countries. Many such projects have been conducted through the World Health Organization (e.g., Bennett, Campillo, Chandrashekar, & Gureje, 1998; Bennett, Jança, Grant, & Sartorius, 1993; Riley & Marshall, 1999; Room, Jança, Bennett, Schmidt, & Sartorius, 1996).

Considering the extent to which cross-cultural differences and similarities in alcohol consumption patterns and their consequences have been explored in anthropological writings, it is surprising that no such explicit studies have been conducted in cross-cultural settings regarding illicit beverages. Data regarding the consumption of such beverages are found in many ethnographic accounts, often in the context of a discussion about "traditional" beverages. For example, in Heath's (2000) book comparing drinking across cultures, he described moonshine production and consumption in a variety of cultures, including Costa Rica, Guatemala, Ireland, Russia, Scandinavia, Tanzania, Ukraine, and the United States, although this is not a particular focus of his work. Similarly, the place of moonshine in alcohol beverage drinking is encompassed in several of the country studies in Heath's 1995 *International Handbook on Alcohol and Culture*. Countries studied include Canada (Cheung & Erickson, 1995); Egypt (Ashour, 1995); Guatemala (Adams, 1995); Honduras (Vittetoe Bustillo, 1995); Iceland (Ásmundsson, 1995); Malaysia (Arokiasamy, 1995); Mexico (Natera Rey, 1995); Poland (Moskalewicz & Zielinski, 1995); Russia (Sidirov, 1995); the United States (Hanson, 1995); and Zambia (Haworth, 1995). Thus, data are available from many cultures on the role of illicit beverages, but such data were not purposively collected around a consistent theme until the cross-national study reported in this volume.

ASPECTS OF ILLICIT DRINKING PATTERNS IN THE SIX COUNTRIES STUDIED

Russia

Samogon is clearly the dominant illicit beverage consumed in Russia, as seen in chapter 3 (Zaigraev). The extent to which it is consumed—vis-à-vis commercially produced beverages such as vodka—has been tracked over time and has been shown to respond to changes in general prohibition policies in the country. The impetus of prohibition policies has varied over time as well. Although many of the increases in *samogon* production and consumption have been connected to governmental policies restricting commercially produced alcohol beverage consumption (e.g., the 1985–1988 antialcohol campaign), some periods of *samogon* increases have been related to perceptions by the general populace that certain commercially produced beverages were dangerous to drink (e.g., "fake vodka" in the early 1990s and the many reported cases of people being poisoned after drinking it).

Russia is a society in which alcohol beverage consumption generally has been a core feature of both routine and special occasions. According to Zaigraev, alcohol consumption is "an inalienable element of the Russian lifestyle." A recent article in *The New Yorker* traces the complex history of vodka over the past 500 years in Russia (Erofeyev, 2002). In their sociological survey of families in three regions, the Russian investigators found that virtually all adults drank alcohol beverages. In addition to the widely recognized position of vodka consumption, imbibing of *samogon* is widespread. Even though alcohol beverage consumption frequency is considerably lower among women than men, about a fourth of the women surveyed drank frequently. The investigators also found that people over the age of 50 drank more frequently than younger age cohorts.

Although pressures resulting in the increase of *samogon* consumption varied, key informants in all three regions studied reported rationales for drinking *samogon* in preference to vodka. People saw vodka as being very expensive, given their limited income. *Samogon* could be produced and purchased much more cheaply than vodka. They also believed that *samogon* produced at home was of a better quality and healthier to drink than vodka.

Regional variation in the proportion of *samogon* consumption compared with other alcohol beverages was examined, but very slight differences between the three regions in the ratio between *samogon* and vodka consumption were found. In two regions (Voronezh and Omsk) *samogon* production had a long historical tradition, and the survey data indicated that it accounted for about 83% of the alcohol beverages consumed. In the Nizhegorod region, 82% of the alcohol consumption reported by participants in the study was *samogon*. In the latter case, the total amount of alcohol consumption was considerably less than in the other two regions; apparently *samogon* brewing is relatively recent in the Nizhegorod region, commencing in the period 1985–1988 during the imposition of strict antialcohol policies.

In rural areas particularly, as pointed out in chapter 3, people have little incentive to cease producing *samogon* because of long-standing traditions and sentiments associated with its production and consumption. These factors need to be seriously considered in attempts to understand and modify patterns of production and consumption of illicit beverages.

In Russia it was found that there were major differences between men and women in the extent to which they consumed *samogon* in comparison with vodka. Women in these surveys reported drinking less than half the total amount of alcohol beverages consumed by the men; however, although the total amount of vodka consumed by men and women was very similar, men drank almost three times as much *samogon* as women. Thus, *samogon* is clearly the overwhelming drink of choice for these men when they drink, whereas vodka is the more common drink of choice among women. Such gender differences would be worth exploring more deeply, especially when considering possible intervention strategies for the consumption of *samogon* in particular. Intervention measures that are effective for men may not be effective for women because there could be different reasons men and women choose to drink *samogon* at particular times and places.

One of the clear messages of the Russian study was that *samogon* is very widely consumed in these regions among adults of all ages and among both men and women. Drinking *samogon* is a social activity in all three communities studied, with approximately 85 to 95% of the drinking occasions over the previous month occurring in social situations. Aging does not deter drinking; in fact, alcohol consumption generally appears to increase as people age, especially after the age of retirement. Furthermore, *samogon* production and consumption do not appear to be abating, but increasing. Reasons for the increase and prospects for a continued rise are especially important to consider, but a particularly interesting reason given is that *samogon* is apparently being used as a currency during economically difficult times. Approximately half the samples of *samogon* that were tested for chemical composition were purchased from people the respondents knew or were produced by the consumers themselves. These features of *samogon* distribution indicate that the substance is deeply imbedded in the socioeconomic lives of Russian people.

From the information provided, it appears that the survey was conducted mainly among respondents from rural areas, even though these areas are in the vicinity of major cities (e.g., Omsk and Nizhegorod). In order to develop effective intervention strategies to curb *samogon* production and consumption, it would be helpful to have similar data on city dwellers, especially with regard to the use of *samogon* as a currency.

The use of *samogon* as a currency of exchange for necessary products and services would fit very well as an anthropological explanation of increased *samogon* use beyond changes in availability and the relative cost of commercially produced alcohol beverages. In addition, anthropologists would very likely stress the importance of *samogon* consumption in terms of its symbolic meaning to people as they consume it. Thus, it would be helpful to study in depth both the apparently "practical" reasons for its production, purchase, and consumption, as well as the more subtle, yet powerful, reasons for continuing to drink *samogon* even if it contains impure additives.

A final question is to identify the most important objectives of prevention/intervention strategies in Russia in reducing the production and consumption of *samogon*. Is it to protect the Russian people better against the impurities that are consumed when they drink *samogon*? Is it to reduce overall alcohol beverage consumption? Is it to prompt a shift from the drinking of *samogon* as a noncommercial product to consumption of vodka, beer, and other commercially produced alcohol beverages? Depending on the most important objectives, quite different approaches would need to be used in prevention and effective intervention.

India

Many contrasts between Russia and India in illicit beverage consumption emerge from the report in chapter 8 (Gaunekar et al.). Although drinking is widespread in Russia, India presents a contradiction between its reputation as a "dry culture" and a venerable cultural and historical association with alcohol beverage consumption. Variation around alcohol beverage consumption predominates in India, a reflection of the extraordinary cultural differences across the country with regard to religion, language, ethnicity, social class, and ethnic groups. At the same time, an anticolonialist attitude prevails across the country, and alcohol beverage consumption is associated in people's minds with British imperialism. India provides an excellent example for anthropological exploration over time and space, as it moved from colonial rule to independence in 1947 and into contemporary times. It is a rich context within which to examine the behaviors and meanings around consumption and abstinence from alcohol beverage consumption. The wide-ranging place of traditional alcohol beverages in Indian society adds to the relevance of an anthropological perspective.

India's phenomenal cultural variation is reflected first and foremost in the broad array of different traditional beverages commonly consumed, in sharp contrast with Russia where the single illicit beverage *samogon* predominates. A collection of typical illicit beverages, which varied depending on the region examined, was identified in the study. Furthermore, class was a critical factor is determining whether either commercially or noncommercially produced alcohol beverages were drunk in particular cultural groups. One widespread feature in India with regard to alcohol beverage consumption is that women rarely drink. This has been a consistent finding in all the general population surveys conducted. For that reason, the Indian study was conducted with men only. In addition, only men who were seen to be "hazardous" drinkers were included. India shows a relatively low incidence of hazardous drinking; among the few surveys carried out to determine hazardous drinking incidence and prevalence, reported rates have ranged from 0.5% to 3.4% of the population. At the same time, the report in chapter 8 suggests that the level of hazardous drinking might be much higher in contemporary India.

In the Indian study, the term "illicit alcohol" was used. The investigators view illicit alcohol as being a subtype of noncommercially produced alcohol beverages in that it is not sold under official governmental controls. The terminology attached to commercially produced, traditional, illegal, and other types of alcohol is quite complex in India and would provide an interesting area of exploration in its own right. This also emerged in earlier reports (e.g., Bennett et al., 1993, 1998; Room, et al., 1996). Realistically, it is difficult to define what is meant by "an illicit alcohol beverage" that would apply to all of India.

Tremendous regional variation exists in India as to legal constraints on alcohol beverage consumption. The study found a sharp contrast between the four regions in terms of prohibitions, formal controls, and taxes. In Goa, on India's west coast, alcohol consumption is liberally permitted, in contrast to the other three regions. This region remained under Portuguese rule until 1961, which provided a strikingly different cultural history compared with the Indian states that were controlled by Great Britain until 1947. Furthermore, the taxes on such beverages imposed by the state in Goa are particularly low. In sharp contrast, Ahmedabad in the state of Gujarat, also located in western India, has a strong formal policy for the control of alcohol beverage consumption. Although strong prohibition policies in Gujarat effectively prevent the purchase and consumption of commercially produced alcohol beverages among most of the population, noncommercially produced beverages are consumed, although under very secretive conditions. In addition to these two highly contrasting regions in western India with regard to alcohol beverage controls generally, two other regions fall in between. New Delhi in northern India evidences open alcohol beverage availability, with formal controls on distribution and costs. The fourth region, a rural one, is located in southern India in the state of Tamil Nadu. In this state, formal alcohol policies are more liberal than in Gujarat, but more restrictive commercially than in either Goa or New Delhi.

Whatever the status of governmental controls on alcohol beverage availability and purchase, it is clear from chapter 8 that illicit beverage consumption is widespread vis-à-vis consumption of commercially produced beverages throughout different regions of India. At the same time, whether such beverages are considered to be illicit depends on the laws pertaining in each Indian state. A particularly interesting twist on this pattern is reported for Goa: In general, there are very few restrictions on producing illicit beverages. *Feni*, produced typically from cashews or coconut, is actually often distilled by breweries and sold commercially in Goa, and receives certain tax advantages.

Specific illicit beverages identified for the other three regions are very different: In Ahmedabad, *thaili*, *tharra*, and *desi daru* are the main beverages; in Vellore in southern India, several beverages are consumed, including *sarayam*; and in New Delhi, *deshi*, a generic term for "country liquor," is common.

Residents of Goa, where Roman Catholicism is common, are religiously more mixed than in the other three regions, which have a large Hindu majority. However, as chapter 8 reports, religion does not seem to be a good explanation for regional variation in overall alcohol beverage consumption or drinking illicit beverages.

Some clear regional differences were found. The ratio of commercial to noncommercial beverage consumption based on the drink histories of hazardous male drinkers reflected a major difference between Goa and the other three regions. Although in Goa the most frequently consumed alcohol beverage was commercially produced (67%), in the other three regions noncommercially produced beverages were more frequently consumed (85% in Vellore, 90% in Ahmedabad, and 91% in New Delhi). Similarly, the proportion of drinking days in which noncommercially produced beverages were consumed contrasts sharply across the four sites. Goan respondents reported consumption of illicit beverages on only 34.7% of drinking days during the study. In contrast, respondents from Vellore, Ahmedabad, and New Delhi reported such consumption on 92.5%, 92.6%, and 89.2% of drinking days, respectively.

Several rationales for relatively frequent consumption of illicit beverages by hazardous-drinking males emerged from this study. The first reason appeared to be the overall availability and cost of commercially produced beverages. There is a large cost difference between the four regions with regard to the two types of alcohol beverages. Religious sanctions may also play a role in these patterns, even though they do not emerge in chapter 8. In Goa, with its relatively greater proportion of Christians to Hindus, it may be easier to consume commercially produced beverages, without need for illicit beverages. Interestingly, hazardous male drinkers in Goa reported consuming fewer total units of alcohol beverages on their drinking days than such drinkers in the other three regions. Solitary drinking, however, was a common feature in all four regions. In India, in sharp contrast to Russia, social occasions that prescribe alcohol beverage consumption are not common.

It is suggested in chapter 8 that formal control policies on production, distribution, and consumption act as major incentives for people to drink noncommercially produced alcohol beverages. Outright prohibition of alcohol beverage distribution in Gujarat appears to contribute to both the production and consumption of illicit beverages. According to Gaunekar et al., attempts to prevent such consumption should combine informed legislative policies with educational programs.

In future studies it would be helpful to explore in depth the relationship between religious practices and drinking practices in specific regions and to see if religious beliefs affect the relative consumption of noncommercially produced beverages in comparison with commercial drinks. Even though women do not drink on average nearly as much as men, it would be interesting to examine the differences between men and women who do drink with respect to informal controls through the family, religion, and the community. An open question would seem to be whether or not hazardous drinking is substantially increasing in India. If so, what kinds of formal and informal controls influence this increase? Finally, because solitary drinking is so common among drinkers in India, an examination of the possible relationship between this manner of drinking and a possible increase in the extent of hazardous drinking would be worth exploring.

Zambia

Although the main illicit distilled beverage referred to in the Zambia study reported in chapter 4 (Haworth) is *kachasu*, this was not the main alcohol beverage consumed in any of the three suburbs studied. In an earlier, larger scale study of HIV infection in Zambia, only 2% of the 4,642 participants reported drinking *kachasu* (Lusaka is being devastated by an HIV/AIDS epidemic). Instead of illicit beverages, the main beverage consumed is Chibuku, which is commercially produced; it is a fermented beverage typically made from sorghum and maize. The data reported from the 140 respondents indicate that *kachasu* was consumed by about 24% of the men and only 9% of the women, in comparison with about 92% of the men and 78% of the women who drank Chibuku during the course of the study. The percentage of respondents who did drink *kachasu* was higher in Kamanga than in the two other, somewhat better off, suburbs.

In the Zambian study, a diary method was used among 120 individuals (88 men and 32 women) who volunteered from 75 households in three locations in the capital city of Lusaka. Two of the suburbs (Kaunda Square and Mutendere) are authorized suburbs, whereas a third (Kamanga) is a shanty town where the inhabitants are relatively poorer than in the other two areas. Daily reports were written regarding each respondent's drinking patterns and context that day and evening. These data included the alcohol beverage consumed, the effect of drinking, and whether drinking led to problems, among other issues. The use of diaries was considered to be a helpful innovation in the study because earlier investigators had had difficulties collecting such kinds of data.

Husbands and wives rarely drank together. Men and women tended to drink in different locations. While approximately 76% of men drank at least part of the time in taverns, only 34% of women did. In contrast, 62% of the men and 94% of the women drank at home sometimes. According to the study compared with earlier research, there has been considerable stability in patterns over the past 25 years regarding with whom one drinks in Lusaka. Although men tended to drink with friends at a tavern or bottle store, women typically drank with relatives or friends at home. At the same time, about 14% of both men and women reported drinking alone. Even though husbands and wives seldom drank together, men might drink with other women.

Because Chibuku was clearly the main alcohol beverage of choice and is a commercially produced beverage, why would people choose to drink *kachasu* some of the time? Another study by Mukuka (2000), which surveyed *kachasu* distillers, concluded that the production and consumption of illicit beverages were the result of poverty. However, the study reported in chapter 4 did not find such a clear distinction between drinkers with relatively greater available money and the others. From the data collected, it appears that there was little price difference between the cheapest commercially produced beverages and many of the illicit beverages. *Kachasu* has a reputation for having a high concentration of alcohol, compared with brewed beverages.

However, when age was taken into account, the study found that people in the 37- to 60-year age group consumed *kachasu* much more than two younger age groups (31% compared with 17% and 13%). These data would be well worth exploring further. Although respondents were asked to identify the alcohol beverage they drank at a particular time, they were not asked to explain why they chose that particular beverage. Therefore, it is difficult to determine what these incentives might be to drink *kachasu*. One possible incentive for the relatively greater consumption of Chibuku is that people believe that they can get intoxicated for less money when they consume Chibuku, compared with other alcohol beverages. From the data collected in the diaries, however, it appears that most respondents were spending beyond their economic means.

A remaining question worth further exploration has to do with the change among middle-aged men and women (37–60 years) to greater consumption of *kachasu*. Because this is a wide age group, it would be helpful to divide it into subgroups and then to collect life histories around consumption patterns. That might clarify how the transition occurs and what the influences are on the choices that people make regarding alcohol beverage consumption. It would also be useful to explore the observation that people seem to spend beyond their means when they pay for alcohol beverages, both commercially and noncommercially produced. How does this happen, and what are its effects on families, and does it lead to problems within the family?

United Republic of Tanzania

In contrast to Russia and Zambia but like India, a substantial majority of the residents of the Tanzanian coastal commercial capital city of Dar es Salaam did not drink alcohol beverages. In chapter 5 (Kilonzo et al.), it is estimated that 77.5% of males and 88.8% of females were abstainers. The high abstention rate of women is also a characteristic shared with most parts of India. Furthermore, in Dar es Salaam those women who drank alcohol beverages drank relatively less frequently than men. At the same time, drinking in other parts of the country was considerably greater. Alcohol beverage consumption patterns in Tanzania place the country among the "medium consumption" nations of the world.

Brewing alcohol beverages has a long tradition in Tanzania. As brewing became commercialized during the 20th century, licensing laws were adopted. According to chapter 5, these laws have had little impact on production and consumption of beers brewed at home. One change is that in earlier times the ritual use of alcohol beverages was emphasized more than becoming intoxicated. Over the past three decades, alcohol beverage consumption has shifted away from mainly ritual use to recreational use. An increase in harmful drinking appears to have occurred as well.

A very substantial percentage of the alcohol beverages consumed in Tanzania is produced noncommercially. According to available survey estimates, about 89% of alcohol beverages drunk are opaque beers that are noncommercially produced. Firm figures are not available for such consumption because there is little reliable documentation. Several types of beers are produced locally or at home. In addition to these beers, a prominent drink is *gongo* or moonshine; this is an illicit spirit produced from fermented pawpaw and sugar.

One concern about both the noncommercially produced beers and *gongo* relates to the amount of impurities and the negative health effects of drinking such toxic beverages. In order to test for impurities, samples were collected of each alcohol beverage consumed in each of the three villages. A separate and more focused sampling was made of *gongo*.

Three villages located approximately 15 km from the city in different directions on the outskirts of Dar es Salaam were the study sites. The villages were selected in order to provide variation in economic base, religion, and types of alcohol beverage consumed. None of the communities was in the upper socioeconomic strata. In the household diaries of adult (that is, aged 12 or older) family members, details about alcohol beverage consumption were recorded. In addition, beverage samples were collected for a chemical analysis of noncommercially produced beverages. Special attention was paid to involvement of family members in traditional brewing of beer. A marked demographic pattern was that younger members of the households drank the most heavily, although their precise ages and whether or not they were drinking homebrew is not clear. The available food crops in the region influenced the choice of alcohol beverage produced: palm wine (*mnazi*) in two communities, honey mead beer (*wanzuki*) in one community, and banana beer (*mbege*) in all three villages. Beer is the most commonly drunk beverage in all three villages. *Gongo*, an illicit distilled spirit, was the next most common alcohol beverage. It was drunk more in the most rural of the three villages, relative to other beverages. When more than one beverage was consumed by participants, beer and *gongo* were by far the most common drinks. With regard to age, being young was associated with heavier drinking.

Consumption of noncommercially produced alcohol beverages (mainly beer) was very common among drinkers in these samples in Tanzania. Most drinkers consumed both commercially and noncommercially produced beverages. Socioeconomic status appeared to be related to the exact choice of alcohol beverage, with the higher status groups drinking mainly commercially produced beverages. Middle status participants consumed both beer and *gongo*. Participants with lower socioeconomic status preferred palm wine. The relative cost of similarly potent beers is important to consider: Drinkers paid about five times as much for a beer that was industrially produced as for a homebrewed beer.

The dominant religion (whether Christian or Islamic) in each of the villages did not appear to be strongly connected to consumption patterns in general. In future studies it would be worthwhile to explicitly explore possible impacts of religious affiliation on drinking patterns.

Following on the point made in chapter 5 about the shift from ritual to recreational drinking, researchers might delve into Tanzania's differing drinking contexts and how they are connected to young people learning to drink. How do the noncommercially produced beverages fit into both ritual and recreational drinking? Because younger men are identified as being a heavier drinking group, it would be interesting to see how young they are when they begin to drink. The high rate of female abstinence from drinking would be interesting to explore, especially in both ritual and recreational drinking contexts. What is happening in Tanzanian society that discourages women from drinking?

Mexico

Mexico exhibits a great deal of variation in types of alcohol beverages typically consumed, as in most of the other five countries, according to chapter 7 (Rosovsky). In addition to traditional beverages that date from before colonial contact, many new beverages have been added to the repertoire of alcohol drinks. The consumption of *pulque*, a fermented beverage, carries special meaning to Mexicans due in part to its roots in precontact culture. Additionally, there is historical evidence to indicate that alcohol beverage consumption did not lead to serious drinking-related problems until after European contact, because there were clear guidelines about acceptable drinking behavior at ceremonials. Following colonial conquest, a sugar-cane distillery was opened in Cuernavaca in the 16th century. Over the centuries since then, various attempts have been made to control production and consumption of alcohol beverages.

During the period since the Second World War, Mexico has experienced accelerated industrialization generally and in the alcohol beverage industry specifically. The Mexican government has attempted to regulate production and consumption of alcohol beverages by law; however, enforcing such policies has proved very difficult. Some of the traditional beverages have been commercially produced and distributed both in Mexico and abroad. *Tequila* is the main example. The most commonly consumed alcohol beverages range from traditional (such as *pulque*) to nontraditional industrialized (such as beer); *tequila* fits into both categories.

Drinking alcohol beverages is part of all religious and social events in Mexico. Yet in estimates of per-capita consumption, noncommercial beverage consumption is not included. Thus, it is suggested in chapter 7 that reported national consumption figures are seriously deflated because they do not include these other beverages. From the surveys conducted, it appears that approximately 15% of the drinkers consume about 78% of the total intake of alcohol beverages. It is estimated that about 34% of the distilled alcohol beverages produced in Mexico are consumed through the noncommercial market.

Even with the wide availability of commercially produced alcohol beverages, *pulque* ranked high among drinking preferences, especially in rural areas. Women in Hidalgo, in particular, appeared to prefer it to beer. As mentioned in chapter 7, *pulque* "is a drink that is there, in their homes or at their neighbors', it is cheap and nutritious, it calms their thirst, and their ancestors drank it."

In a recent study in the Mexican state of Chiapas, anthropologist Christine Eber (1995) explored the changing views of alcohol beverage consumption as community members experienced the Zapatista movement. According to Eber, both women and men became political activists in a movement that rejected alcohol beverage consumption as a positive national symbol. Support for such abstinence came from Protestant churches. It would be interesting to see whether such abstinence movements in various parts of Mexico are conceptualized in the same way for both commercially produced and noncommercially produced alcohol beverages. How does an abstinence movement regard *pulque,* given its long-standing position as a traditional substance? After all, "their ancestors drank it." The study also emphasized the importance of religious affiliations and gender differences in abstinence and recovery efforts. It would be helpful to collect life history data, in particular, in order to elucidate such questions.

Brazil

As in Mexico, alcohol beverages were drunk in Brazil during pre-Columbian times (see chapter 6, Vaissman). The beverage that has remained very popular over the centuries is *cachaça*, which is a drink distilled from sugar cane. Although beer is the most commonly consumed alcohol beverage in Brazil, *cachaça* remains very popular, whether legally and or illicitly produced. In general alcohol beverages are widely available, and regulations on production and consumption are limited.

Early in the Portuguese occupation, the government outlawed the production and consumption of *cachaça*. Ironically, the taxes subsequently imposed on *cachaça* in the mid-18th century produced sufficient funds to help reconstruct the Portuguese capital city of Lisbon following a devastating earthquake in 1755.

The noncommercial production of *cachaça* is most common in the state of Minas Gerais. Although the Ministry of Agriculture controls its production, it is estimated that only 10% of the approximately 8,000 *cachaça* distilleries in Minas Gerais are registered. Because of the low prices of homemade *cachaça* and its high alcohol content, least advantaged members of the society purchase and consume it. According to chapter 6, there is no longer any stigma associated with drinking *cachaça*. Furthermore, its sale does not require a special license.

From the data collected from the 91 participants, homemade *cachaça* was the second most frequently consumed alcohol beverage (following beer, which was overwhelmingly the most popular), with wine a close third. Homemade *cachaça* was most commonly consumed in bars, restaurants, or *biroscas*, rather than in the home. Drinking *cachaça* did not depend on special occasions; drinkers of *cachaça* most often drank alone or with friends or colleagues.

Interestingly, although there are few constraints in Brazil on the production and consumption of alcohol beverages, it is illegal to be drunk in public. Drinking for the sake of drunkenness was not common. From epidemiological research in Brazil, it is clear that drinking alcohol beverages typically begins early in life, and few people abstain from drinking. At the same time, Brazilians do not typically drink in binges. In general, drinking alcohol beverages—especially beer and *cachaça*—is an integral part of everyday life, whereas informal controls from the family and the wider society have a major impact on drinking patterns.

According to a 1999 household survey, approximately 48.8% of men and 23.6% of women consume alcohol beverages. Thus, as in most of the other five countries, men are more likely to drink than women. These data are not consistent, however, with the conclusion that few people abstain from drinking. If the 1999 survey were accurate, it would appear that the majority of both men (over half) and women (three-fourths) do not drink alcohol beverages. That finding would merit further exploration in order to ascertain the overall drinking rates among adults in Brazil with greater certainty.

An intriguing notion that emerges from the Brazil study is the negative value placed on being drunk, or at least on drunken behavior. It appears from chapter 6 that Brazilians have adopted a "Mediterranean" style of sustained heavy drinking that is maintained throughout a large part of the day and evening, with relatively little intoxication. Just how is intoxication controlled? Participant observation of drinking behavior and discussions with both men and women about how they think they control the extent of their own drinking would be interesting. What are their views of people who do become intoxicated and what are their explanations for such "aberrant" behavior?

COMPARISONS ACROSS THE SITES

This section presents a comparison of data from the six country studies with regard to certain aspects of noncommercially produced alcohol beverage consumption that are anthropologically relevant. Because the methods of data collection were not consistent across countries, the section begins with a brief description of the methods used at each study site. Although all the investigators from the six sites explored the status and effects of illicit beverage consumption, there is considerable variation in the types of data available. This poses certain difficulties in making comparisons across the six countries. Despite that variation, information is available on these topics for each country: identification of the dominant illicit beverages in each culture; contrasts in drinking patterns among men and women; general availability of illicitly produced alcohol beverages; concerns about toxic substances in illicit beverages; drinking to intoxication as a desired objective; patterns of solitary drinking; and abstention as a common pattern. These specific issues fit within an anthropological framework and are relevant to understanding the effects of illicit beverage consumption.

Methods of Data Collection

The study in Russia (chapter 3) contained a sociological survey of men and women in the context of their families in three contrasting regions of the country. In addition, a chemical analysis was conducted of 81 samples of *samogon* from each of the three regions in order to test for impurities.

In contrast, key informant interviews, focus groups, and AUDIT questionnaires were used in four distinct regions of India (chapter 8). The interviews and focus groups were aimed at general information about differences in noncommercial alcohol beverages common to the four regions and patterns of availability and usage of those beverages. AUDIT questionnaires were collected from potentially hazardous male drinkers in each of the four regions, and some 50 individuals who returned high scores for hazardous drinking were recruited for the study to provide daily data on their drinking patterns.

In Zambia (chapter 4), the diary method was used among 88 men and 32 women who volunteered from 75 households in three locations in the capital city, Lusaka. Two of the suburbs studied (Kaunda Square and Mutendere) were authorized suburbs, whereas a third (Kamanga) was a shanty town where the inhabitants were poorer than in the other two areas. Daily reports were written regarding each respondent's drinking patterns and context that day and evening. The data collected included alcohol beverages consumed and the effects of drinking, among other issues. Ostensibly the diary method results in reports of higher alcohol beverage consumption than do other methods of data collection.

In Tanzania (chapter 5) the investigators collected household diaries. An interesting method of sampling was used in selecting the households. In each village, from the list of 10-cell leaders, three were selected by lottery. Within each of these three cells all 10 households were recruited. Thus, there were 30 households representing each village, or a total of 90; because all adults aged 12 years or older were included in the study, a total of 199 household members participated. Several components were included in data collection: interviews with the heads of the households regarding drinking patterns of all household members; intermittent visits to the households in order to monitor the diary keeping; measurement of the vessels used to drink alcohol beverages; general field notes on the diary-keeping process; and observational field notes regarding drinking behavior observed while there. The investigators found that with the diary method, higher alcohol beverage consumption was reported than with the AUDIT questionnaire.

In Mexico (chapter 7), diaries were collected from 51 households, 25 from three neighborhoods in Mexico City and 26 from Hidalgo. Men and women over the age of 15 years participated. One family member entered the information for the entire family in the diary.

The diary method was also used in Brazil (chapter 6). A convenience sample was selected in São João de Meriti in the state of Rio de Janeiro, the most densely populated urban area in Brazil. Ninety-one individuals over the age of 15 and representing 77 families participated in the study. Diaries of drinking behavior were maintained by these individuals over the course of 6 weeks. The AUDIT questionnaire was also administered to one or more family members before selection of the sample.

Dominant Illicit Beverages

It is not difficult to identify particular noncommercially produced alcohol beverages that are most regularly consumed in each culture. In Russia, *samogon*, a "strong" (30–60% alcohol content) beverage distilled from agricultural products such as sugar beets and various fruits, is clearly identified. Various illicit beverages are noted for India: *feni*, which is distilled from cashews or coconuts; *toddy,* which is fermented from flowers of palm or coconut trees; urrack, which is distilled from cashews; and *arrack*, which is distilled from products such as wheat. Cultural and botanical diversity have a major impact on the illicit beverage of choice in India. The main illicit beverage consumed in Zambia is *kachasu*, which is distilled from sorghum and maize. *Gongo* or moonshine in Tanzania is a fermented spirit from pawpaw and sugar. Tanzanians also consume several home-brewed legal beers and wine. In Mexico, *pulque* is fermented from the juice of the maguey cactus, and it is the primary illicit beverage consumed. *Cachaça*, which is distilled from sugar cane, is the main illicit alcohol beverage in Brazil. In short, a primary noncommercially produced alcohol beverage is found in five of the six countries, whereas in India a variety of such beverages is consumed.

Availability of Noncommercially Produced Beverages

Although there is some variation in the degree to which illicit beverages are available in different parts of Russia, *samogon* is generally readily available. In India there is much greater variation in availability of either commercially or noncommercially produced beverages. In three of the regions studied, both types are reportedly freely available, while in the fourth (Ahmedabad), all alcohol beverages are considered illicit because of the general prohibition on alcohol beverages. Although not the most commonly consumed alcohol beverage in the three neighborhoods of Lusaka studied, in Zambia the main illicit beverage consumed is *kachasu.* It was drunk by respondents from all three neighborhoods. In Tanzania, there was widespread availability of noncommercially produced alcohol beverages, including opaque beers and *gongo*. In fact, most people reported that they preferred such types of drinks, and policies to control the production and consumption of either locally or home produced beer or *gongo* were not very effective. As far as selection for consumption was concerned, it appears that the boundary between commercially and noncommercially produced alcohol beverages was quite vague. Participants seemed to move back and forth between alternatives with few worries about whether what they were consuming was "legal" or not. Availability and cost appeared to be much more important considerations. Traditional beverages of both legal types were very popular with participants. In Mexico noncommercially produced beverages are readily obtainable, especially *pulque*. Alcohol beverages, both legal and illicit, are widely available in Brazil and by and large are not regulated by formal alcohol control policies. *Cachaça*, a distilled beverage, is produced through both commercial and noncommercial channels, and it is not difficult for people to obtain illicit *cachaça*.

Drinking by Men and Women

In the Russian sociological survey, both men and women were included, while only men were encompassed among the key informants, focus groups, AUDIT data collection sample and participants in India. The decision to include only men in India is a reflection of the infrequency of drinking by women. Both men and women were included in the Zambian study; while husbands and wives seldom drank together, women did drink. The Tanzanian data indicated that in two of the three villages studied women consumed slightly more units of alcohol beverages than the men. At the same time, women spent less for those drinks than did the men. The source of this discrepancy would be interesting to identify: since similar proportions of men and women were brewers themselves (about 14% of the participants), drinking their own brews cannot be the only explanation. In both Mexico City and Hidalgo women reported drinking less often than men. In Mexico City beer was the preferred beverage for both men and women, while in Hidalgo both beer and *pulque* were popular with the men and *pulque* with the women. In Brazil, men are about twice as likely to drink alcohol beverages as women (in one study 48.8% versus 23.6%).

Drinking to Become Intoxicated

According to chapter 1, drinking in Russia characteristically involves consuming large amounts of potent alcohol beverages, frequently without eating food. In the study itself, it is difficult to determine that the respondents regularly drank to become intoxicated. Many of the respondents denied becoming highly intoxicated and avoided questions relating to alcohol-related problems. Even so, the majority of those who answered reported such problems, including two-thirds who experienced severe hangovers. Based on the study data and other sources, it can be concluded that drinking occasions in Russia often result in intoxication for many participants. This issue is not explicitly addressed in the Indian report. In Zambia the predominant style of drinking described was imbibing with the intent to become intoxicated. Interestingly, drinking the illicit beverage *kachasu* was reported to be associated with drinking to intoxication more than when other illicit beverages were consumed. The data from Tanzania do not address this question one way or the other. In Mexico both men and women often drank to become intoxicated, though the pattern pertained more to men than to women. In contrast, strong informal controls through the family and friends discourage drinking to intoxication in Brazil. Public intoxication is frowned on, even though alcohol beverage consumption is not.

Solitary Versus Social Drinking

In Russia, most of the drinking was social drinking. In contrast, in India respondents described drinking alone very often. Drinking occasions described in the Zambian study were mainly social activities. Apparently solitary drinking is not common in Tanzania. In Mexico both men and women drank with other people most of the time, whether they drank in public places or at home. In the Brazilian sample, over 50% of the units of *cachaça* consumed over the 6 weeks of the study were drunk when the study participant was alone. In contrast, only 11% of the units of beer reported to have been consumed during this time period were drunk in a solitary fashion.

Abstention

Abstention was found to be relatively common in India and Tanzania. It does not appear to have been as common in the other four countries.

Toxic Substances as an Issue

In Russia concerns about toxic substances that have harmful effects on consumer's health were clearly significant, reflected in part by the fact that a chemical analysis of 81 samples from across the three regions was carried out. This was also a focus of the Indian study, and analyzed samples evidenced very high amounts of ethers and acids. Additives to illicit beverages were identified, and some—such as battery acid and ammonia—were clearly toxic. In Zambia it was reported that people believe that *kachasu* is the strongest alcohol beverage available, whether commercially or noncommercially produced, and that toxic ingredients are added. Impurities are a concern in Tanzania with regard to both the locally produced beers and the illicitly distilled *gongo*. Samples of both types of beverages were analyzed and found to contain impurities, especially *gongo*. *Gongo*, in fact, is reported sometimes to contain sufficient methyl alcohol to cause blindness or death. Its alcohol content is very high, compared with the beers or palm wine. In Mexico, the toxicity of *pulque* and *aguardiente* were analyzed, with the former evidencing relatively low levels of ethanol and methanol and the latter evidencing very high levels of ethanol. In Brazil, the toxity of *cachaça* was not investigated. (Chatper 11 contains more details on these analyses.)

SOME CONCLUDING "ANTHROPOLOGICAL" QUESTIONS

When the consumption and effects of consumption of illicit beverages across different cultures are examined, certain questions emerge. For example, what is the status of prohibition of alcohol beverage consumption generally within specific cultures? What types of explicit prohibitions against consumption of alcohol beverages not produced by the recognized alcohol industry exist? To what extent do people in particular cultures see illicit beverages in the same category as commercially produced alcohol beverages? Do they view them as being as potent or more potent? What political and economic pressures exist from inside and outside the country to encourage or discourage the consumption of noncommercially produced alcohol beverages? With regard to legal constraints on illicit beverage consumption, how clear are such laws and to what extent does the average person understand them? Similarly, what is the society's track record in enforcing these prohibitions or restrictions? What is the relationship currently and in the past between traditions of alcohol beverage consumption and commercially produced beverages? How are estimates of per capita consumption made, and how do they take into account consumption of illicit beverages? What is the threat to health from drinking noncommercially produced alcohol beverages in comparison with commercially produced ones? What kinds of toxic substances add to the potential for negative effects on health?

In exploring such questions, ethnographic and quantitative methods used in combination are very effective. Observations of people's drinking behaviors, including consumption of different categories of alcohol beverages, can lead to questions to be explored in interviews, life histories, even focus groups. In this cross-national study, the diary approach provided a very useful means of tracking consumption over time and under different circumstances. That method, used jointly with survey and ethnographic approaches, holds considerable promise for better understanding the place of illicit beverage consumption within a given society. Multicountry studies such as those described here can do much to help identify patterns across cultures as well as more idiosyncratic behaviors and belief patterns.

REFERENCES

Adams, W. R. (1995). Guatemala. In D. B. Heath (Ed.), *International handbook on alcohol and culture* (pp. 99–109). Westport, CT: Greenwood.

Ames, G., Delaney,W., & Janes, C. (1992). Obstacles to effective alcohol policy in the work place: A case study. *British Journal of Addiction, 87*, 1055–1069.

Arokiasamy, C. V. (1995). Malaysia. In D. B. Heath (Ed.), *International handbook on alcohol and culture* (pp. 163–178). Westport, CT: Greenwood.

Ashour, A. M. (1995). Egypt. In D. B. Heath (Ed.), *International handbook on alcohol and culture* (pp. 63–74). Westport, CT: Greenwood.

Ásmundsson, G. (1995). Iceland. In D. B. Heath (Ed.), *International handbook on alcohol and culture* (pp. 117–127). Westport, CT: Greenwood.

Bennett, L. A., & Ames, G. M. (Eds.). (1985). *The American experience with alcohol: Contrasting cultural perspectives.* New York: Plenum.

Bennett, L. A., Janča, A., Grant, B. F., & Sartorius, N. (1993). Boundaries between normal and pathological drinking: A cross-cultural comparison. *Alcohol, Health, & Research World, 17*, 190–195.

Bennett, L. A., Campillo, C., Chandrashekar, C. R., & Gureje, O. (1998). Alcohol beverage consumption in India, Mexico, and Nigeria: A cross-cultural comparison. *Alcohol, Health, & Research World, 22*, 243–252.

Cheung, Y. W., & Erickson, P. G. (1995). Canada. In D. B. Heath (Ed.), *International handbook on alcohol and culture* (pp. 20–30). Westport, CT: Greenwood.

Eber, C. (1995). *Women and alcohol in a highland Maya town: Water of hope, water of sorrow.* Austin: University of Texas Press.

Erofeyev, V. (2002, December 16). Letter from Moscow: The Russian God. *The New Yorker*, pp. 56–63.

Gefou-Madianou, D. (Ed.). (1992). *Alcohol, gender, and culture*. New York: Routledge.

Hanson, D. J. (1995). The United States of America. In D. B. Heath (Ed.), *International handbook on alcohol and culture* (pp. 300–315). Westport, CT: Greenwood.

Haworth, A. (1995). Zambia. In D. B. Heath (Ed.), *International handbook on alcohol and culture* (pp. 316–327). Westport, CT: Greenwood.

Heath, D. B. (Ed.). (1995). *International handbook on alcohol and culture*. Westport, CT: Greenwood.

Heath, D. B. (2000). *Drinking occasions: Comparative perspectives on alcohol and culture*. Philadelphia: Brunner/Mazel.

Hill, T. W. (1985). On anthropology and ethnography: A problem in the history of anthropology. *Current Anthropology, 26*, 282–284.

Hunt, G., & Barker, J. C. (2001). Socio-cultural anthropology and alcohol and drug research: Towards a unified theory. *Social Science & Medicine, 53*, 165–188.

Kunitz, S. J., & Levy, J. E. (1994). *Drinking careers: A twenty-five year study of three Navajo populations*. New Haven, CT: Yale University Press.

Leland, J. (1976). *Firewater myth: North American Indian drinking and alcohol addiction* (Monograph 11). New Brunswick, NJ: Rutgers Center of Alcohol Studies.

Levy, J. E., & Kunitz, S. J. (1974). *Indian drinking: Navajo practices and Anglo-American theories*. New York: Wiley.

MacAndrew, C., & Edgerton, R. (1969). *Drunken comportment: A social explanation*. Chicago: Aldine.

Marshall, M. (Ed.). (1979). *Beliefs, behaviors, and alcoholic beverages: A cross-cultural survey*. Ann Arbor: University of Michigan Press.

Marshall, M. (Ed.). (1982). *Through a glass darkly: Beer and modernization in Papua New Guinea* (Monograph 18). Boroko: Institute of Applied Social and Economic Research.

Marshall, M., Ames, G. M., & Bennett, L. A. (2001). Anthropological perspectives on alcohol and drugs at the turn of new millennium. *Social Science & Medicine, 53*, 153–164.

Moskalewicz, J., & Zielinski, A. (1995). Poland. In D. B. Heath (Ed.), *International handbook on alcohol and culture* (pp. 224–236). Westport, CT: Greenwood.

Mukuka, L. (2000). *A baseline study of the extent to which illicit alcohol (kachasu) is used and abused by periurban communities in Lusaka and Mazabuka*. Lusaka: Centre for Social Policy Studies, University of Zambia.

Natera Rey, G. (1995). Mexico. In D. B. Heath (Ed.), *International handbook on alcohol and culture* (pp. 179–189). Westport, CT: Greenwood.

Riley, L., & Marshall, M. (1999). *Alcohol and public health in 8 developing countries*. Geneva: World Health Organization.

Room, R. (1984). Alcohol and ethnography: A case of problem deflation? *Current Anthropology, 25*, 169–191.

Room, R. (2001). Intoxication and bad behaviour: Understanding cultural differences in the link. *Social Science and Medicine, 53,* 189–198.

Room, R., Janča, A., Bennett, L.A., Schmidt, L., & Sartorius, N. (1996). WHO cross-cultural applicability research on diagnosis and assessment of substance use disorders: An overview of methods and selected results. *Addiction, 91*, 199–220.

Sidorov, P. I. (1995). Russia. In D. B. Heath (Ed.), *International handbook on alcohol and culture* (pp. 237–253). Westport, CT: Greenwood.

Strunin, L. (2001). Assessing alcohol consumption: Developments from qualitative research methods. *Social Science & Medicine, 53*, 215–276.

Vittetoe Bustillo, K. W. (1995). Honduras . In D. B. Heath (Ed.), *International handbook on alcohol and culture* (pp. 110–116). Westport, CT: Greenwood.

Chapter 10

Key Economic Issues Regarding Unrecorded Alcohol

Eric Single

The cross-national studies presented in the first part of this volume were intended to explore the availability of data and implications of noncommercial or otherwise poorly recorded sources of alcohol. It is clear from the case studies that there is enormous variability regarding levels of unrecorded consumption. In some countries locally produced, largely noncommercial alcohol accounts for a relatively minor share of total alcohol consumption (e.g., Tanzania—see chapter 5). However, in many countries locally produced alcohol accounts for the lion's share of consumption, whereas commercially produced alcohol represents only a small portion of total consumption (e.g., Russia—chapter 3).

The purpose of this chapter is to discuss economic aspects of noncommercial alcohol and other types of unrecorded alcohol production. The chapter begins with a discussion of the various sources of unrecorded alcohol, alternative strategies for measuring unrecorded consumption, and the underlying factors that influence the relative volume of unrecorded alcohol. The discussion then focuses on the consequences of unrecorded alcohol for the epidemiological monitoring and economic analyses of alcohol consumption, with particular attention to the impact of significant unrecorded consumption on estimates of the economic costs of alcohol misuse. The chapter concludes with some policy implications of unrecorded alcohol consumption, including the need for more research to inform policymakers on optimal levels of alcohol controls and the development of effective harm reduction measures aimed at minimizing problems associated with noncommercial alcohol.

TYPES OF UNRECORDED CONSUMPTION AND MEASUREMENT ISSUES

There are several types of alcohol production that often are not reflected in official statistics on alcohol. These include both commercially produced alcohol and noncommercial sources of production. Commercially produced but poorly recorded or undocumented sources of alcohol include: (1) commercially produced alcohol that is sold illegally; (2) legally imported quantities of alcohol for personal consumption; (3) illicit importation of commercially produced alcohol; and (4) by-products of commercially produced alcohol and commercially produced nonpotable alcohol. In addition, there are at least three types of noncommercial alcohol: (5) alcohol illicitly produced on a large scale; (6) local small-scale production of alcohol outside the formal economic system; and (7) production of alcohol at home for personal use.

Each of these types of alcohol presents special challenges for estimation and measurement:

Commercially produced alcohol that is sold illegally. It appears that there has never been a published study that estimates the volume of illegal sales of commercially produced alcohol. This is possibly because illicit sale of commercial products is rarely considered to be a significant type of unrecorded consumption. In countries with exceptionally good recording systems for production and sales, it might be possible to estimate the volume of commercially produced alcohol that is sold illegally by comparing production to sales data. However, the volume of inventories would have to be taken into account, or at least would have to be assumed over time to be roughly equivalent to the prior recording period. Another potential source of data for estimating the amount of commercial alcohol sold illegally would be criminal justice data on the number of persons arrested or convicted for illegally selling alcohol. This, however, is probably more a reflection of enforcement activity than a good measure of illicit production. Finally, in situations where illicit selling is commonplace and where there are no sanctions against purchasers, a crude estimate of the extent of illegal selling of commercial alcohol could possibly be obtained from general population surveys.

Legally imported quantities of alcohol for personal consumption. In some countries, the volume of alcohol that is imported by consumers may have an appreciable impact on overall levels of consumption. This applies particularly to tourist-based economies and in regions where trade barriers against alcohol imports have been relaxed (as in the European Union). The number of border crossings is an important consideration. In Europe and some other developed economies, the volume of travelers is high, and legal importation of small quantities could affect overall consumption. In Canada, for example, there are more than 1 million border crossings each day in a population of 30 million (Single & Giesbrecht, 1979). In such cases, it is possible to combine data on the influx and outflow of travelers with survey data on the typical amounts of alcohol imported or exported within personal limits to estimate the net impact on total alcohol consumption. However, in the only two studies that appear to have estimated this figure (Mäkelä, Österberg, & Sulkunen, 1981; Single & Giesbrecht, 1979), it was concluded that the importation of small amounts of alcohol for personal use had no appreciable impact on overall consumption levels. Furthermore, in many developing countries there are relatively few international travelers, so this type of unrecorded consumption is not likely to be significant.

Illicit importation of commercially produced alcohol. The illicit importation of commercial alcohol has been a problem at times when there are wide discrepancies between the price of alcohol in neighboring jurisdictions. Up to the past decade, this was the situation in Canada and the United States. Lower prices for alcohol in the United States led to a problem of illicit importation of U.S. alcohol into Canada (Offer, 1987). It is difficult to measure illegal activities such as this. Changes in the number of persons charged with illicit importation may be used as a crude measure of trend (but not level) if there is no change in enforcement activity. Survey data might also produce a rough estimate of the frequency of illicit importation. In any case, it is generally believed that, for most countries, this is not a significant source of unrecorded consumption.

By-products of commercially produced alcohol and commercially produced nonpotable alcohol. The final type of unrecorded consumption of commercially produced alcohol concerns the use by alcohol-dependent persons of commercial by-products and nonpotable alcohol, such as cooking wine, or methanol-based products. The quantity involved is generally considered to be too small to significantly influence overall levels of consumption. Estimations of the magnitude of by-product and nonpotable alcohol consumption would have to rely on expert informants (e.g., social welfare specialists working with chronic drunks, police) or emergency-room admission data on alcohol poisonings.

Illicit production. Unlicensed production may contribute significantly to alcohol consumption levels in some countries as indicated in the chapters on Brazil, India, Mexico, Russia, and Zambia. The quantity can only be roughly estimated from survey or diary data provided by the consumers. Such data should also include information on the alcohol content of the beverages produced. Data on the number of persons detected and sanctioned for violating laws against unlicensed production may also be indicative of trends (if levels of enforcement remain relatively constant).

Local, small-scale production of alcohol beverages outside the formal economy. This category refers to legal production of local alcohol beverages for a local market by small-scale producers (often home producers). Although legal, such production is largely unregulated, untaxed, and unrecorded. Thus it is essentially the same in terms of its economic impacts as illicit production. It is also estimated in much the same way by survey or diary data.

Alcohol produced legally by the consumers themselves for home consumption. Although generally a legal activity, home-produced alcohol is also similar to illicit production in that it is largely unregulated, untaxed, and unrecorded. Again, the magnitude of home production can only be roughly estimated using data on raw materials and survey or diary data (see, for example, Single & Giesbrecht, 1979). In some countries, these data should be supplemented with data on assisted home production at "U-brews" and "Make-your-own-wine" shops if these businesses significantly increase the amount of home production.

FACTORS INFLUENCING THE VOLUME OF UNRECORDED ALCOHOL

There are as yet an insufficient number of studies estimating the volume of unrecorded consumption to warrant a thorough meta-analysis of the factors underlying the amount of unrecorded consumption in a society. However, the country case studies presented in this volume indicate a number of potential determinants. Clearly, high prices and the limited availability of commercially produced alcohol are important influences. Countries with high levels of locally produced, unrecorded consumption are invariably those where commercially produced alcohol is very expensive or otherwise difficult for most consumers to obtain.

Another factor is the extent to which small-scale producers can obtain relatively easy entry into the alcohol market. This depends on the ready availability of natural resources and raw materials needed for production and the low levels of technology required to produce local alcohol products, involving skills not necessarily learned in a formal educational system. It is noteworthy that the volume of noncommercial alcohol production in Zambia is influenced by the amount of annual rainfall (see chapter 4). The availability of labor is another factor that reduces barriers to entry into alcohol production. As noted in chapter 3 in this regard, it is the older Russians who appear to produce the most local alcohol, probably because older persons have more time available to make these beverages.

A favorable public attitude toward noncommercial alcohol represents another variable that influences levels of unrecorded alcohol production. Finally, where there are laws governing small-scale alcohol production at the local level, corruption and lack of enforcement can undermine the impact of such controls. For example, bottle stores are not permitted to sell alcohol for on-premise consumption in Zambia, yet many such stores provide seats and even entertainment for patrons (see chapter 4).

CONSEQUENCES OF UNRECORDED CONSUMPTION

Unrecorded alcohol has significant economic consequences. First, the various categories of unrecorded alcohol create the same consequences—positive and negative—that result from recorded, commercial production of alcohol beverages, although not necessarily to the same degree. As with commercial alcohol, there are clearly social benefits from locally produced noncommercial alcohol and other forms of unrecorded alcohol production. To the extent that unrecorded alcohol contributes to the number of persons consuming alcohol at low levels, there are also the same cardiovascular and other health benefits that derive from commercial alcohol consumed in moderation (see, e.g., English et al., 1995, for a meta-analysis of positive and negative impacts of alcohol use). Small-scale noncommercial production also brings certain economic benefits to local economies, providing employment and income (often supplemental income) to producers and lower priced alcohol to consumers. By the same token, noncommercial alcohol and other forms of unrecorded consumption entail much the same adverse consequences as does the use of commercial alcohol. These include higher risk for a variety of chronic diseases, higher risk of accidents, and social problems such as unemployment, productivity loss, marital discord, and alcohol-related violence. As is well documented in cost-of-illness studies, these adverse consequences can have significant impacts on the economy, increasing health care costs, forcing governments to invest in prevention, research, and training, and causing significant productivity losses (e.g., Collins & Lapsley, 1996; Rice, Kelman, Miller, & Dunmeyer, 1990; Single, Robson, Xie, & Rehm, 1998). As with commercial alcohol, the quality of life for high-volume consumers of noncommercial alcohol may suffer as the result of excess spending on alcohol.

In addition to the consequences that flow from alcohol use per se, noncommercial alcohol and other forms of unrecorded consumption tend to carry some additional and more negative consequences. Noncommercial alcohol has been implicated as a frequent cause of alcohol poisoning. The results of toxicological tests of noncommercial alcohol in the case studies in this volume indicate that, in many countries, the likelihood of poisoning from noncommercial alcohol is not necessarily higher than the likelihood of poisoning from commercially produced alcohol. Nonetheless, if only because of the ethanol in these beverages, noncommercial alcohol clearly contributes to overall levels of alcohol poisoning. Moreover, where local noncommercial production of alcohol is high, there is significant loss of revenues for governments and loss of business for commercial producers. To commercial producers, local small-scale alcohol production represents unfair competition because it is less regulated and generally untaxed.

For economists, alcohol researchers, managers of commercial alcohol firms, and government policymakers, unrecorded alcohol represents a serious impediment to research and planning. It leads to underestimation of true levels of alcohol consumption that can undermine economic planning and creates extreme difficulties for epidemiological monitoring of alcohol use and research on alcohol-related problems. Thus, improved estimates of the extent of noncommercial and other forms of unrecorded alcohol would be of benefit to a wide variety of parties, including the alcohol industry, public health researchers, and governments.

THE IMPACT OF UNRECORDED ALCOHOL ON ECONOMIC ANALYSES

It is extremely difficult to assess the economic impact of alcohol use where the volume of noncommercial alcohol is significant. Cost estimation studies and other forms of economic analyses only concern economic consequences for the legitimate market economy. Cost estimation studies do not generally attempt to measure costs arising from the economic disruption to legitimate business enterprises caused by large-scale illicit alcohol production and distribution. Indeed, in Russia there are situations where money is scarce and noncommercial alcohol actually becomes an alternative to the official currency (see chapter 3). To measure the full economic ramifications of noncommercial alcohol in such situations, an extensive and demanding economic framework such as a general equilibrium model would need to be used. There has never been an economic analysis of noncommercial alcohol using a general equilibrium model.

Studies estimating the economic costs of alcohol are severely restricted in situations where a significant portion of consumption occurs outside the formal market economy. As stressed earlier, unrecorded alcohol leads to underestimation of consumption. This in turn leads to underestimation of alcohol-attributable morbidity and mortality. Deaths and hospitalizations attributable to alcohol use are generally estimated by combining information on the relative risk of using alcohol at different levels for various causes of disease and death with information on the number of persons consuming alcohol at levels associated with higher relative risks (English et al., 1995). These two pieces of information—relative risk and prevalence—are combined to generate the etiologic fractions, or the proportion of cases for a particular cause of disease or death that can be reasonably attributed to alcohol use. When there is significant use of noncommercial alcohol, the prevalence of persons consuming at levels associated with a higher relative risk is underestimated, thus causing serious underestimation of alcohol-attributed morbidity and mortality. Many of the costs in economic cost studies flow from these morbidity and mortality estimates. In particular, health care costs and productivity losses, which constitute the two largest economic costs of alcohol use in most cost estimation studies, depend on estimates of the number of hospitalizations and deaths caused by alcohol use. Thus, the existence of significant unrecorded consumption invariably leads to serious underestimation of economic costs caused by alcohol use.

In addition, some economic costs may well be disproportionately higher as a result of consumption of noncommercial alcohol. Health care costs will be influenced by higher rates of alcohol poisonings. Where the use of noncommercial alcohol is so widespread that it significantly reduces life expectancy, as appears to be the case in Russia today (see chapter 3), there will be a consequent increase in productivity losses due to premature mortality. Enforcement and prevention costs may also be disproportionately higher in societies with high rates of use of noncommercial alcohol.

CONCLUSIONS: THE POLICY IMPLICATIONS OF UNRECORDED ALCOHOL

The implications of the foregoing discussion can be usefully separated into two categories: implications of unrecorded, commercially produced alcohol and those of noncommercial, local production. With regard to commercial alcohol, in many countries there is a need for better enforcement of existing laws regulating the sale of commercial alcohol and illicit importation of commercial products. With regard to the legal importation of small quantities of commercial alcohol for personal consumption, it is unclear if this represents a serious problem in any country at the present time. If this were deemed to be of significance in a number of countries, greater international consensus would be required as to appropriate controls over the importation of commercial alcohol for personal use (such as agreements on duty-free limits). With regard to commercial by-products and commercially produced nonpotable alcohol, in those countries where this represents a significant problem, special regulations may be required to reduce alcohol poisoning from these sources, such as the obligation to account for by-products and restrictions on access to products such as cooking wine, over-the-counter medications with high alcohol content, and methanol-based products.

The portion of unrecorded alcohol that appears to cause the most problems is noncommercial alcohol, including homemade products for personal use, illicit production, and legal, small-scale, local production of alcohol. Precisely because such production is largely outside the formal economy, noncommercial alcohol is especially difficult to control or regulate. The regulation of noncommercial alcohol, as with any social policy, entails dual and at times conflicting goals: The aim is to reduce problems associated with the production and use of noncommercial alcohol while at the same time minimizing the costs and adverse consequences that result from efforts to control production and use. The lack of regulation of noncommercial alcohol leads to loss of revenue and greater problems, but at the same time excessive controls could lead to greater noncommercial production. Thus, any attempt to better regulate noncommercial alcohol must necessarily involve a balancing of goals.

Perhaps what is most needed is research indicating more precisely the optimal levels of alcohol controls so that noncommercial alcohol is kept to a minimum and the harm that can arise from alcohol consumption (whether commercial or noncommercial in origin) is also minimized. A starting point is case studies such as those presented in this volume, describing the social history of small-scale local alcohol production and use in a variety of settings. Eventually, when sufficient evidence emerges from such case studies, it may be possible to develop an empirically testable theoretical model of the key determinants of noncommercial alcohol. The model would be likely to involve the availability of commercial alternatives (in both price and convenience); the availability of technology and raw materials for noncommercial alcohol; public acceptance of noncommercial alcohol; price of noncommercial alcohol relative to disposable income; and the extent and effectiveness of government interventions such as treatment and prevention programs. Once these key determinants are identified, it may be possible to develop an empirical database with information from many countries to determine optimal levels of controls that would minimize noncommercial production and at the same time effectively reduce alcohol-related problems. Unfortunately, that goal is far from being achieved. At this point, even robust estimates of the extent of noncommercial alcohol production and use for most countries are lacking.

In any case, there is an immediate need to develop better interventions to ameliorate the adverse consequences of noncommercial alcohol use without leading to greater levels of consumption. The quick and easy response to such problems is to advocate greater restrictions and increased enforcement. But excessive regulation could lead to even greater noncommercial production and thus only exacerbate the situation. New and inventive interventions and prevention initiatives are needed that recognize the benefits that noncommercial alcohol offers to consumers, accepting the fact that such alcohol use will continue. Perhaps most promising in this regard is the emergence of "harm reduction" measures for alcohol in the recent past (Single, 1997). Rather than attempting to reduce levels of consumption, harm reduction measures aim at reducing the adverse consequences that may result from drinking. As applied to the problem of noncommercial alcohol, harm reduction measures might include educational programs to educate consumers on how to deal with alcohol poisoning, and other such programs for noncommercial producers on how to reduce the health risks associated with their products. Licensing requirements could be introduced for the production of certain products that are currently unlicensed, with provision for training producers and distributors to recognize and deal with problems that might arise from the use of these products. Governments could also offer free testing of toxicity levels of noncommercial products.

It must be recognized, however, that the problem of noncommercial alcohol is greatest in those countries that can least afford such prevention programming. Indeed, in more extreme situations such as societies where noncommercial alcohol accounts for the lion's share of total alcohol consumption, it might be argued that it would be of net benefit to public health to reduce taxes or restrictions on commercial products in order for them to compete more effectively with noncommercial products. Given the risks involved in such a policy, however, it would be imprudent to subscribe to this position until there is more compelling evidence. The magnitude of the costs of noncommercial alcohol would need to be estimated and compared with the potential costs associated with lower taxes and restrictions on commercial alcohol. At present, as mentioned, even the magnitude of noncommercial production is unknown. Clearly, the next steps are to conduct further case studies, to promote the use of standardized measuring techniques, and to share information on potentially effective preventive measures, in order to develop a progressively more extensive database to better understand and effectively deal with the problem of noncommercial alcohol.

REFERENCES

Collins, D., & Lapsley, H. (1996). *The economic costs of drug abuse in Australia in 1988 and 1992.* Canberra, Australia: Commonwealth Department of Human Services and Health.

English, D., Holman, D., Milne, E., Winter, M., Hulse, G., Codde, G., Bower, C., Corti, B., de Klerk, C., Lewin, G., Knuiman, M., Kurinczuk, J., & Ryan, G. (1995). *The quantification of drug caused morbidity and mortality in Australia, 1992.* Canberra, Australia: Commonwealth Department of Human Services and Health.

Mäkelä, K., Österberg, E., & Sulkunen, P. (1981). Drink in Finland: Increasing alcohol availability in a monopoly state. In E. Single, P. Morgan, & J. de Lint (Eds.), *Alcohol, society and the state, Vol. 2: The social history of control policy in seven countries* (pp. 31–60). Toronto, Canada: Addiction Research Foundation.

Offer, S. (1987). *Report of the Advisory Committee on Liquor Regulation/Rapport du Comité Consultatif sur la Réglementation des Alcohols.* Toronto, Canada: Queen's Printer.

Rice, D., Kelman, S., Miller, L., & Dunmeyer, S. (1990). *The economic cost of alcohol and drug abuse and mental illness 1985* (DHHS Publication No. ADM 90-1694). San Francisco: Institute for Health and Ageing, University of California.

Single, E. (1997). The concept of harm reduction and its application to alcohol: The 6th Annual Dorothy Black Lecture. *Drugs: Education, Prevention and Policy, 4*, 7–22.

Single, E., & Giesbrecht, N. (1979). The 16 percent solution and other mysteries concerning the accuracy of alcohol consumption estimates based on sales data. *British Journal of Addiction, 74*, 165–178.

Single, E., Robson, L., Xie, X., & Rehm, J. (1998). The economic costs of alcohol, tobacco and illicit drugs in Canada, 1992. *Addiction, 93,* 983–998.

Chapter 11

Chemical Composition, Toxic, and Organoleptic Properties of Noncommercial Alcohol Samples

Vladimir Nuzhnyi

The history of alcohol consumption coincides with the history of humanity. From the earliest times, alcohol beverages have been an integral part of everyday life and culture and are often classified as nutritional products, despite their negative effects on the population.

Alcohol is the only legal substance with pronounced psychotropic effects in most countries and is less dangerous than other available psychotropic substances of plant or synthetic origin. At the same time, we can conclude that people need to use psychotropic substances from time to time and will drink to correct or change psychological and physiological conditions.

Alcohol beverages produce many changes in the human body, more than other psychoactive substances of either natural or artificial origin. It is possible, therefore, to study these products from different angles—as nutritional products, as substances that cause addiction and that can produce toxic effects, and as substances that have a positive impact on health.

Nutrition and other properties of alcohol beverages are determined first of all by ethanol. Ethanol is an excellent nutritional substrate with an energy value of 7.1 calories per gram. When used in small and moderate amounts, it is almost completely oxygenated in the liver, and hence it is of value as a food.

Besides ethanol, many alcohol beverages contain a lot of other components that originate from plants, from oak containers the beverages are stored in, or from the fermentation process. Other components are deliberately added to beverages. Grape wine and beer have the most components. Drinking these beverages in moderation can fully or partly satisfy human body needs for some nutrition substances. For example, two glasses of wine a day will provide a person with a 40% increase in flavinoids (Waterhouse & Frankel, 1993), while drinking 0.5 liters of beer daily will satisfy the need for nicotinic acid and folic acid (Nuzhnyi, 1997).

According to the concept of structured information in nutrition science, food's biological value is determined to a considerable extent by the amount of minor components. Although these have sometimes been considered to be of small importance and even ballast, they significantly increase the amount of structural information that a person receives with food (Brechman & Kublanov, 1983). Compared with other alcohol beverages, vodka has minimal structural and informational potential. According to this criterion, it can be classified as a refined product like white sugar, refined vegetable oil, or white flour. On the contrary, fermented and probably distilled alcohol beverages, including traditional homemade ones, can play an important role for people who do not get a lot of essential substances with their food.

Discussion of alcohol's negative effects and toxicity is very topical, especially in the Russian Federation. Alcohol-related medical problems in Russia are marked by a high proportion of alcohol-related poisonings in total mortality. For example, within the last decade Russian mortality related to alcohol and alcohol surrogate varied between 18 and 38 cases per 100,000 people—much higher than in other European countries (Ivanets & Koshkina, 2000). Researchers have also found a high level of alcohol-related morbidity. Some 40% of patients of working age in Moscow have diseases that are directly or indirectly related to alcohol (Ogurtsov, 1989).

The high level of acute, alcohol-related mortality is usually explained by high average per-capita consumption, by Russia's "Nordic" style of drinking, and by the widespread consumption of noncommercial alcohol, including *samogon* and surrogates with high toxicity. In terms of the number of deadly alcohol poisonings, Russia has always been different from other countries. At the beginning of the 20th century, the Russian rate of lethal poisonings was five times that of other European countries, even though average per-capita consumption was lower (Bechterev, 1913; Kulapina, 1998). These data suggest that biological causes can determine differences in a population's resistance to alcohol's toxic effects.

A great number of experimental, clinical, and population studies have been carried out on the toxic effects of alcohol and alcohol beverages. All have clearly proven that the toxic effects of different alcohol beverages are mostly determined by ethanol rather than by other components. Toxic effects are observed with daily drinking of alcohol in amounts higher than 0.5–0.8 grams of pure ethanol per kilogram of body weight. Some components and admixtures of nonalcohol origin can modify the toxic effects of ethanol. For example, the polyphenolic components of grape wine can limit the toxic effect of alcohol. On the contrary, other components common in different alcohol beverages (including ethyl carbamate, polycyclic carbohydrates, furfurol, methanol, biogenic amines, and phytoestrogens) can increase ethanol's toxic effects, as can ingredients that are sometimes deliberately added to beverages (ethylene glycol, oxalic acid, some plant extracts, etc.).

The strongest impact on ethanol metabolism comes from alcohol dehydrogenase isoenzyme ADH2 and by aldehyde dehydrogenase isoenzyme ALDH2. Allel isoforms ADH2-2 and ALDH2-2 have abnormal activity that leads to accumulation of acetaldehyde in the body in the case of exogenic alcohol consumption. In particular, homozygote and heterozygote forms of ADH2 (ADH2-2/2 and ADH2-1/2) have 10 and 20 times the activity of ordinary alcohol dehydrogenase and can cause hyperproduction of the powerful toxin. Homozygote and heterozygote forms of ALDH2-2 (ALDH2-2/2 and ALDH2-1/2), with activity close to zero, delay further oxidation of acetaldehyde, causing almost a 20-fold increase of its blood concentration (Edenberg & Bosron, 1997; Meier-Tackmann, Leonhardt, Agarwal, & Goedde, 1990).

Among those who abuse alcohol, these genotypes increase the risk of alcohol poisoning, lung disease, pancreatitis, hepatitis, cirrhosis, and cancer of the mouth, pharynx, esophagus, colon, and lungs. They also decrease the amount of alcohol that can cause poisoning (Suchodolova, Strachov, & Chundoeva, 2000; Yamauchi, Maezawa, Toda, & Sakurai, 1995; Yin, Liao, Chen, & Lee, 1992; Yokoyama, Muramatsu, Omori, & Ishii, 1999). At the same time, however, those who have genes with abnormal activity have a decreased risk of alcohol dependency (Boras, Coutelle, Rosell, & Parres, 2000). The frequency of alleles ADH2-2 and ALDH2-2 is rather high among many Asians (Japanese, Chinese, Koreans, Vietnamese, Thais, Filipinos, and Malaysians). The frequency equals 85 to 20% for ADH2-2 and 30 to 0% for ALDH2-2. In Western and Central Europe prevalence of these alleles is rather low: 5.4 to 0% and 0.9 to 0%, correspondingly (Boras et al., 2000; Goedde & Agrawal, 1992). According to the results of pilot studies, Russians have a 41% incidence of the ADH2-2 allele, a middle ground between Asians and Europeans (Ogurtsov, Garmash, Miandina, & Moiseev, 2001).

It is likely that the preferences of different ethnic groups for different alcohol beverages are determined not only by cultural traditions, but also by the prevalence of the ADH and ALDH genotypes within the group. In particular, Japanese researchers have established that carriers of the ALDH2-2 gene prefer beverages with a low concentration of alcohol. On the other hand, individuals with the homozygote genotype ALDH2-1 prefer beverages with high alcohol concentration (Ishibashi, Harada, Fujii, & Ishii, 1999).

These substances can also have their own direct effects on organs and tissues. Unfortunately, information about the biological effects of different types of alcohol beverages is mostly based on an analysis of their chemical composition and on the results of population surveys. Experiments and toxicological studies in this field are rarely done.

The last two decades have seen an information explosion concerning positive health effects of alcohol. These effects are primarily related to ethanol itself and to polyphenol components. It has been proven that some alcohol beverages, especially red grape wine, prevent ischemic heart disease (Waterhouse & Frankel, 1993) and produce stress-reducing and antiradiation effects (Nuzhnyi & Prichozhan, 1996). Other positive effects, however, have not yet been proven.

There are very few studies of the chemical composition of noncommercial alcohol and its toxicity. One clinical study carried out in India of 144 patients with alcohol-related hepatitis and liver cirrhosis found no differences in the severity and course of disease among patients who used homemade alcohol and those who used high-quality, foreign industrial liquors (Sarin, Malhotra, Nayyar, & Broor, 1988).

Another experimental study, in which pregnant or lactating mice and Syrian golden hamsters were given a solution of noncommercial alcohol, demonstrated that the sample of homemade alcohol used had a carcinogenic effect (Zariwala, Lalitha, & Bhide, 1991). Later, the same group of authors showed that this carcinogenic effect might be related to its influence on cytochrome Ð-450 activity and to the level of glutaminate in the liver (Zariwala, Kayal, & Bhide, 1993). Another recent publication of Indian scientists studied the ability of homemade *toddy* to affect lipid metabolism among pregnant rats. It found that, compared to an ethanol solution of the same concentration, *toddy* causes a more pronounced change of biological synthcsis and lipid exchange and leads to the development of more pronounced alternative changes in the liver. It is assumed that these differences are conditioned by the impact of the nonethanol components of alcohol beverages (Lai, Kumar, Suresh, & Indira, 2001).

METHODS OF ANALYSIS

The methods used in the Russian Federation are first described in some detail, and reference is then made to methods used in those other countries whose reports are considered in this chapter.

Alcohol content in the samples was analyzed by the density of solutions and by alcohol measuring tables.

Gas chromatography–flame ionization detector (GC-FID). A Hewlett-Packard 5890 gas chromatograph (GC) with a flame ionization detector (FID) was used in the analysis. The temperature of the injection port was 240° C, while the detector temperature was 220° C.

Gas chromatography–mass spectrometry (GC-MSD). In this analysis, we used the GC-MSD system (Hewlett-Packard), including the mass selective detector (MSD) HP-5973 and HP-6890 gas chromatograph. The energy of ionization was 70 eV. The signal was recorded in a total ion current (TIC) mode in the range of m/z 29–400. Temperature in the injection port was 240° C, with a temperature of 180° C in the interface to the mass spectrometer. Analyzed substances were identified using the NIST98 standard library of mass spectra.

Conditions of Chromatographic Separation

Two types of chromatographic condition of separation on capillary columns of different length were utilized in the analysis.

Method 1

Separation was performed on capillary column HP-FFAP (free fatty acid phase, ether of polyethylene glycol and terephthalic acid). The column size was 25 m × 0.32 mm, and thickness of the film was 0.52 μm.

The following temperature program was used: 45ºC (2 min); 10ºC/min; 190ºC. The pressure at the head of column was 6 psi. An injection sample volume was 1 μl with split 1/16.

Method 2

A HP-FFAP column of 50 m length, 0.20 mm internal diameter, and 0.33 μm film thickness was used in GC-MSD, with the temperature program of 50ºC (2 min); 10ºC/min; 190ºC (20 min).

The flow rate of the carrier gas (helium) was 1 ml/min. Injection sample volume was 1 μl with 1/16 split.

INTERNAL STANDARD

Cyclohexanol was used as an internal standard and was added to the analyzed samples up to a concentration of 28 mg/L (30 μl of 0.1% cyclohexanol solution in ethanol were added to a 1-ml sample).

Quantitative analysis was performed using an internal standard. We used cyclohexanol as an internal standard, as it is an atypical component of alcohol beverages (Savchuk, Brodski, Formanovski, & Rudenko, 1999; Savchuk, Vlasov, Appolonova, & Grigorian, 2001). The detection limit for cyclohexanol was 0.1 mg/L in gas chromatography and 0.05 mg/L in GC-MSD (in a TIC mode).

In GC-FID analysis, substances were identified by retention time (RT) of standard mixture components. Standard mixtures (in ethanol) containing components of alcohol beverages (higher alcohols, ethers, fatty acids, aldehydes, and ketones) and some light solvents were prepared (Savchuk et al., 1999).

We used GC-FID for preliminary analysis, and GC-MSD was used for confirmation of the quantitative analysis results and for confirmation of component identification. The HP-FFAP column with high GC resolution was used in GC-MSD analysis. The GC resolution of columns was estimated by the completeness of the separation of a pair of substances, benzaldehyde/octanol-1, with close retention time.

RT and mass spectrum of standard mix components in case of chromatographic overlapping peaks of analyzed substances was performed with GS-MSD confirmation analysis.

The following overlapping or incomplete separations of GC-FFAP peaks were observed:

1. Dimethyl ether of ethylene glycol/isopropyl alcohol/ethanol/benzene.
2. Ethylene glycol/ethyl decanoate/butyric acid.
3. Propionic acid/2,3-buytanediol.
4. Methanol/isopropyl acetate/methyl-ethyl ketone.
5. Ethyl formate/methyl acetate.
6. Octanol-1/benzaldehyde.

We conducted chemical analyses of alcohol beverages in India, Mexico, Tanzania, and Zambia using routine chemical methods that are used in laboratories to control the beverages' quality and adherence to official health standards. Requirements for the quality and safety of alcohol differ by country. This is why results are presented in universal form and units of measurements: ethanol content in volume percent (% vol.) and content of other components in milligrams per liter of beverage.

Chemical analysis of 12 samples of homemade *gongo* from Tanzania (four samples from each studied region) was carried out using the GC-FID method, with a Varian 3300 model chromatograph. Conditions of analysis were FID and employing column Porapak at temperature range 130–200°C at change rate 2°C/min and a flow rate of 35 ml/min.

A standard preparation 5% of pure ethanol, methanol, propanol, and butanol was mixed at equal ratios of 1:1:1:1. Two microliters of mix were injected into the GC. Two microliters of the bulk sample were then injected into the GC where the concentration of the intended alcohols was calculated with respect to a specific standard chromatogram.

Methodological details of the analyses carried out in Mexico and India are not presented here, because this is a routine procedure used in laboratories of the health services of these countries.

TABLE 11.1. Chemical Composition of *Gongo* From Tanzania

	Settlement		
Substance	Chamazi	Kimara	Gesaulole
Ethanol, % vol.	28.2 (25.1–31.8)	21.1 (14.8–29.4)	24.9 (19.1–32.7)
Methanol, mg/L	0.4 (0–0.16)	16.65 (0–66.6)	0
Propanol, mg/L	1.93 (0.32–6.84)	0.08 (0–0.40)	0
Butanol, mg/L	0	0	0

Note. In this table and other tables, the contents of volatile admixtures (methanol, propanol, buthanol, etc.) and nonvolatile admixtures (lead, copper, zinc, etc.) are given in mg/L beverage.

THE CHEMICAL COMPOSITION OF BEVERAGES FROM TANZANIA, MEXICO, INDIA, AND ZAMBIA

The locations from which beverages were obtained are described in the various country chapters. In Tanzania, the ethanol content in beverages varied between 14.8% and 32.7% vol. Butanol was not present in any of the studied samples, and the sample from Gesaulole settlement had no methanol or propanol. One notable fact is that methanol and propanol content in samples from the other two settlements were 10–100 times lower than in most known distilled alcohol beverages, including industrial ones. Only methodological defects that took place during analysis can explain this unclear phenomenon.

The ethanol concentration of *gongo* differs significantly from most homemade Tanzanian beverages. According to the previous study's results, other types of alcohol beverages (*komoni, kibuku, mbege, mnazi,* beer, *konyagi,* and *amarula*) contain between 3.2% and 8.9% vol. of ethanol (Kilonzo & Pitkanen, 1992).

Table 11.2 presents the results of a chemical analysis of the distilled beverage *aguardiente* from Mexico. Samples were obtained in the three settlements of Hidalgo State where the study was carried out.

TABLE 11.2. Chemical Composition of *Aguardiente* From Mexico

	Settlement		
Substance	Santa Teresa	Santa Maria	San Antonio
Ethanol, % vol.	46.4	35.1	46.6
Aldehyde as acetaldehyde, mg/L	58.32	20.36	32.15
Higher alcohols as amyl alcohol, mg/L	249.96	0	0
Methanol, mg/L	0	0	0
Furfurol, mg/L	0	0	0
Lead, mg/L	0	0	0
Copper, mg/L	0	29.66	0
Zinc, mg/L	0	0	0

TABLE 11.3. Ethanol and Methanol Content in *Pulque* From Mexico

	Settlement		
Substance	Santiago de Anaya, Hidalgo	Hermosillo, Hidalgo	Puerto, Hidalgo
Ethanol, % vol.	5.2	5.0	4.4
Methanol, mg/L	37.00	44.76	35.00

Ethanol content in *aguardiente* is almost two times higher than in Tanzanian *gongo*. Aldehyde content is higher than in typical industrially produced distilled alcohol beverages. As in the previous case, attention should be drawn to the absence of methanol in all the beverages as well as to the absence of higher spirits in two samples. It is most likely that in this case, a methodological defect was also present.

Table 11.3 presents the ethanol and methanol contents of the low-alcohol beverage *pulque*, made with the use of cactus fermentation. Samples were obtained in three settlements of Hidalgo State. The relatively low methanol content has to be pointed out.

Contents of ethanol and of other volatile and nonvolatile components in five samples of distilled and fermented beverages from India are presented in Table 11.4.

All the samples from India had a relatively low content of aldehydes and a very high content of acids and ethers. The content of other components did not exceed established norms.

Results of chemical analysis of minor components in noncommercial alcohol beverages used in Zambia are presented next. Table 11.5 includes results of a chemical analysis of six samples of distilled alcohol beverage *kachasu*. Analysis has been made using routine chemical procedures that are used for quality control of alcohol production.

The quantities of volatile components (acetaldehyde, methanol, ethyl acetate, and higher spirits) in all samples are typical for distilled beverages. Moreover, content of methanol and higher spirits in these samples is even lower than in many industrial beverages. Sample 1 from Mutendere showed abnormally low content of acetaldehyde (6.9 mg/L). Content of copper in all the *kachasu* samples is also at a level that is typical for this category of alcohol beverage. However, the content of lead in four of the samples is heightened and is 1.0 and 1.1 mg/L (according to Russian Federation sanitary standards, the maximal acceptable concentration of lead is 0.3 mg/L). Content of nonvolatile and volatile acids in most of the samples is within the acceptable limits. Only in one *kachasu* sample is the quantity of free acids strongly increased

TABLE 11.4. Content of Alcohol and Other Components in Samples From India

	Samples of beverages					
Substance	Cashew *feni* (A)	Cashew *feni* (B)	Coconut *feni* (A)	Coconut *feni* (B)	*Urrack*	Vellore illicit
Ethanol, % vol.	46.3	42.6	24.8	36.3	18.3	12.2
Aldehydes as acetaldehyde, mg/L	40.7	63.9	17.4	34.8	23.2	11.5
Higher alcohols as amyl alcohol	Passes the test	Passes the test	Passes the test	Passes the test	Passes the test	Passes the test
Test for methanol	Negative	Negative	Negative	Negative	Negative	Negative
Total acid as tartaric acid, mg/L	1512.2	719.9	2339.9	792.1	3671.9	601.1
Volatile acid as acetic acid, mg/L	691.3	299.5	869.7	414.5	1670.4	255.0
Ethers as acetate, mg/L	804.7	499.0	1048.3	542.3	917.2	497.1
Copper as Cu, mg/L	Passes the test	Passes the test	Passes the test	Passes the test	Passes the test	Passes the test
Test for aluminum chloride	Negative	Negative	Negative	Negative	Negative	Negative

Note. Quantitative indicators of alcohol production in India are provided in terms of grams per 100 liters of absolute alcohol. According to the quality requirements of India, the maximum limit counted in other units is 208.3 mg/L for aldehydes; 115.7 mg/L for total acid; 46.3 mg/L for volatile acid; and 231.5 mg/L for ethers. Normal value for ethanol: 19.9–42.8% vol.

(6.0 g/L as acetic). This is a defect of the beverage that occurred during fermentation or storage. Thus, according to the results of the chemical analysis, the majority of analyzed samples of *kachasu* must have good organoleptic properties. It is not likely that the toxic effects of these samples are different from those of industrially produced distilled alcohol beverages.

Table 11.6 shows the results of the analysis of six wine samples. Content of acetaldehyde in these samples varies within the range of 35.0–104.0 mg/L; within 22.0–176.0 mg/L for methanol; within 11.5–167. 5 mg/L for ethyl acetate; and within 149.7–322.8 mg/L for higher spirits. By comparison, the content of acetaldehyde in industrially produced wines varies between 2.0 and 120.0 mg/L; between 20.0 and 350.0 mg/L for methanol; between 20.0 and 200.0 mg/L for ethyl acetate; and between 150.0 and 600.0 mg/L for higher spirit (Pasini, 1992). As with most *kachasu* samples, the content of lead in the

TABLE 11.5. Chemical Composition of Kachasu Samples

	Settlement					
	Kamanga		Mutendere		Kaunda Square	
Substance	No. 1	No. 2	No. 1	No. 2	No. 1	No. 2
Acetaldehyde, mg/L	115.0	229.0	6.9	91.0	173.0	104.0
Methanol, mg/L	198.0	42.0	88.0	52.0	264.0	88.0
Ethyl acetate, mg/L	261.2	2291.5	241.6	147.3	230.7	259.2
Higher alcohol as amyl alcohol, mg/L	523.6	243.1	289.6	242.9	271.4	311.0
Total acid as tartaric, g/L	1.9	3.9	2.4	4.2	3.9	3.6
Volatile acid as acetic, g/L	0.3	0.7	0.4	0.7	6.0	0.6
Copper, mg/L	0.8	0.8	0.7	0.7	0	0.7
Lead, mg/L	0	1.1	1.0	1.0	0	1.0

Note. The study (Tables 11.5 and 11.6) used the method of Horwitz (1975).

four samples of wine is heightened, and the content of copper is within the normative limits. The content of nonvolatile acids in all the samples, except one, is above 15 g/L (as tartaric). Such a concentration is considered to be the maximum for industrially produced young wines. According to the Russian Federation standard, the mass concentration of nonvolatile acids has to be within the limits of 4.0–8.0 g/L (as tartaric), and mass concentration of volatile acids should not exceed 1.5 g/L (as acetic) for all groups of wines. On the whole, analysis suggests that the majority of noncommercial wine samples from Zambia have low organoleptic properties due to their high acidity.

TABLE 11.6. Chemical Composition of Noncommercial Wine Samples From Zambia

	Settlement					
	Kamanga		Mutendere		Kaunda Square	
Substance	No. 1	No. 2	No. 1	No. 2	No. 1	No. 2
Acetaldehyde, mg/L	69.0	61.0	104.0	35.0	35.0	35.0
Methanol, mg/L	22.0	79.0	44.0	44.0	44.0	176.0
Ethyl acetate, mg/L	111.5	116.6	134.6	159.9	167.5	157.0
Higher alcohol as amyl alcohol, mg/L	231.5	322.8	309.1	260.5	149.7	320.8
Total acid as tartaric, g/L	2.4	14.2	1.3	20.2	14.9	16.9
Volatile acid as acetic, g/L	0.10	1.30	2.00	1.48	0.90	1.80
Copper, mg/L	0.72	0	0.68	0.72	0	0.78
Lead, mg/L	0	0.86	0	0.86	1.73	0.82

CHEMICAL COMPOSITION, TOXICITY, AND ORGANOLEPTIC PROPERTIES OF NONCOMMERCIAL ALCOHOL SAMPLES FROM RUSSIA

Analysis of Chemical Composition in Samples From the Russian Federation

We collected samples of homemade beverages from the rural populations of three Russian Federation regions (Nizhnyi Novgorod, Voronezh, and Omsk) where the sociological study was also carried out. We numbered all samples and provided explanatory notes about where and when each of them had been taken. We collected 81 samples of homemade beverages in 0.25-ml glass or plastic containers.

We also analyzed the chemical composition of such liquors as whisky, Cognac, and grappa, made with the use of a distillation technique. According to the samples' descriptions, all tested alcohol beverages were made from sugar; a few samples were made from molasses.

In descriptions of spirit-making technology, one can see the following stages: (1) sugar syrup preparation, (2) syrup processing by yeast, and (3) single distillation, using a self-made distiller with disposal from the head and tail fractions (although this is not necessary). Double distillation was rare. Sometimes, beverages were infused with extracts from plants or other raw materials (oak rind, lemon rind, honey, tea leaf, prunes, or cedar nut). Alcohol samples after extraction were slightly or intensively colored, from light yellow to light brown. Some of the colored samples were opaque and left sediment after a while.

Consumers made about half of the samples themselves, and the other half were bought from neighbors or friends.

The average alcohol content in tested samples was 41.5%. In samples from Nyzhniy Novgorod the average was 40.5% (24.2%–50.6%); from Voronezh, 45.1% (39.6%–62.2%); and from Omsk, 39.4% (16.5%–47.0%).

As to chemical composition, alcohol samples from different regions made with the use of different techniques were similar to industrially distilled spirits (Table 11.7). Nevertheless, homemade alcohols had some distinguishing features.

All of the tested samples had low methanol concentration (0.48–28.6 mg/L), except for one sample that showed an abnormally high methanol concentration, 655.3 mg/L. By comparison, ethyl alcohol "Extra," used to make vodka, can contain methanol up to 240 mg/L. In such strong alcohol beverages as whisky and brandy, methanol content is about 200–1000 mg/L (Nuzhnyi, 1995; Shatirishvili, 1998). However, homemade beverages contained a lot of ethyl acetate, a substance similar to ethanol in its volatility. Therefore, low methanol content is related not to distillation conditions, but to the composition of raw materials and the fermentation process.

TABLE 11.7. Comparison of the Chemical Composition of Homemade Beverages and Factory-Made Spirits: (A) Homemade Beverage (Samogon) (Average Figures for 80 Samples), (B) Russian-Made Cognac (5 Years), (C) Cognac "Henessy V.S." (France), (D) Whisky "Cutty Sark" (Scotland), (E) Grappa (Italy)

Substance	Concentration, mg/L				
	A	B[a]	C	D	E
Acetaldehyde	61.2	88.3	36.9	30.4	174.0
Substance RT 2.25	1.0	0	0	0	0
Acetone	3.8	1.7	0	0	2.0
Ethyl formate	1.0	2.2	4.1	0	2.4
n-Butyraldehyde	0	2.7	0	0	2.2
Ethyl acetate	165.5	1132.4	184.5	134.0	281.5
Methanol/isopropyl acetate	19.0	177.5	166.2	61.0	1621.5
Isovaleric aldehyde	2.1	0	0	0	2.7
Diacetyl	2.4	24.4	0	0	9.9
Substance RT 5.05	0.4	0	0	0	0
sec-Butanol	11.9	44.1	0	0	34.5
Propanol-1	170.1	124.3	137.6	282.7	247.1
Isobutanol	1128.7	162.1	660.0	348.4	299.4
Isoamyl acetate	8.0	13.6	3.7	8.8	9.6
Butanol-1	14.5	23.3	2.4	3.9	11.9
Limonene	1.0	0	1.5	0	0
Isoamyl alcohol	1443.2	613.8	1529.6	373.9	666.2
Substance RT 9.92 (acetoin)	18.3	108.4	0	0	3.8
Hexanol, ethyl lactate	9.8	250.1	67.8	3.7	78.0
Substance RT 11.82	1.2	0	0	0	0
Acetic acid	266.8	494.3	150.7	78.5	63.9
Furfurol	1.0	5.5	12.4	3.0	3.0
Propionic acid	2.6	10.1	0	0	4.3
Benzaldehyde	21.6	0	0	0	0
Isobutyric acid	89.6	3.7	0	0	4.8
Butyric acid	3.6	23.8	0	0	36.2
Ethyl decanoate	5.4	0	33.6	22.8	0
Diethyl succinate	0.1	0	0	0	0
Isovaleric acid	28.2	14.1	10.9	0	11.1
Valeric acid	1.1	22.1	21.0	0	5.3
1,2-Propanediol	0.1	7.5	0	0	15.6
Phenylethyl alcohol	46.7	8.5	24.9	12.7	6.3
Dodecyl alcohol	5.6	0	0	0	0

[a]Low quality according to organoleptic characteristics.

A high concentration of isovaleric acid is typical for all tested homemade beverages, but not for other alcohol beverages manufactured using a distillation or rectification method.

It is also typical that although homemade spirits made from sugar have more acetone than methyl acetate or ethyl formate, factory spirits made from sugar do not.

Most samples contained limonene in concentrations of 0.5–1.0 mg/L, but in some samples the concentration reached 20–30 mg/L. In some cases, this could be related to extraction from lemon rind. Some factory-made beverages used to contain limonene, but they no longer do.

The quality of homemade beverages was evaluated according to the following criteria: (1) the lactic acid/alcohol fermentation ratio, (2) the presence or absence of an oxidizing process during fermentation, and (3) the presence or absence of large amounts of admixtures that are typical for the last fraction of distillate.

The lactic acid/alcohol fermentation ratio was measured by estimating the ethyl lactate concentration. Concentration up to 200 mg/L (measured in pure alcohol, out of water) does not influence the alcohol beverage's quality. But the substances that go together with ethyl lactate during lactic acid fermentation do decrease the quality of cognac and spirits (Sibachov, 2001).

Oxidizing processes were controlled according to diacetyl concentration in the samples. In wines and cognacs these substances have an acetone smell in flavor. There are therefore standards for diketones concentration in beer: no more diketones than 0.1 mg/L (Kunze, 1998). A high content of acetone and ethyl lactate usually coincides with a high content of acetic acid.

Distillation quality was evaluated according to the concentration of last fraction components: acetic acid, propionic acid, isobutyric acid, and phenyl ethanol. According to these three criteria, all tested samples were divided into three groups (Table 11.8). The first group (about 50% of the samples) contained small quantities of admixtures, compared with other distilled liquors (such as whisky and Cognac). Homemade beverages from this group may be considered high-quality alcohol beverages.

Homemade spirits from the second group (about 30% of samples) contained many nonvolatile components, including fatty acids and higher alcohols that are typical for last fraction (fusel oil). Such a low quality could be related to the infringement of the distillation process, meaning that the process was terminated later than necessary. Homemade beverages from this group were marked as having low organoleptic characteristics.

The third group consisted of samples (17.5%) with atypical chemical composition. Samples from this group included enormously high concentrations of diacetyl acetone and ethyl lactate; many components from the etheric and aldehyde fraction (acetone, ethyl acetate, acetaldehyde); or a significant quantity of last fraction substances (isobutanol, phenylethyl alcohol, acetic acid, isovaleric acid, and others). Some samples from this group contained a relatively high concentration of limonene. One sample contained benzaldehyde, which regularly appears in almond oil as an odorant. The third group sample ("D" in Table 11.8) was one of the worst in that its chemical composition showed evidence of defects of the fermentation and distillation process.

Analyzing the materials just presented, we would like to reiterate that quality standards for alcohol production in different countries vary widely. In addition to quality indicators, many countries have normative indicators that are used for toxicity evaluation and safety control.

Since 1996, the safety of alcohol beverages in the Russian Federation has been evaluated by the Health Service Laboratories, which bases its evaluation on content of methanol, lead, arsenic, cadmium, and mercury (copper and iron were once counted as well). For beer, additional control of nitrosamines and radionucleotides (cesium-135 and strontium-90) is conducted. The Russian Federation has long been concerned with substances formed during fermentation that cannot be separated later in the process of ethanol rectification. Besides methanol, these substances include aldehydes, superior alcohols, ketones, and ethers, all commonly considered to increase ethanol's toxic potential. To address the issue, vodka producers in Russia have recently started to use technologies that allow maximum separation of these admixtures from ethanol.

The set of indicators that measures beverage safety is a subject of severe criticism. Results of specially conducted studies and expert observation from the Russian Federation have demonstrated that the content of heavy metals in wine, beer, and distilled alcohol beverages is comparable to that of other nutritional products, and that they rarely exceed established norms. The possibility of chronic intoxication with heavy metals from wine or distilled beverages exists only in the case of massive technogenic pollution of territories used for growing agricultural products. Finally, there is no proof of any relationship between alcohol-related diseases and an excess of mineral chemicals in alcohol beverages (Nuzhnyi & Prichozhan, 1996).

Methanol content must also be analyzed to ensure safety. Methanol is the admixture most difficult to separate during the process of distillation and rectification of ethanol. Organoleptic features of methanol are difficult to distinguish from ethanol. Methanol's interaction with ethanol therefore becomes one of the main causes of lethal poisonings. The methanol content in alcohol beverages that were made by fermentation of apples, plums, pears, and other raw materials that contain pectin may reach significant values of up to 2000 mg/L and more, especially when pectinolytic preparations were used. The concentration of methanol as it exists in most alcohol beverages does not influence acute toxicity of consumed ethanol. In higher concentrations (1000 mg/L and more), depending on the dose, methanol increases the psychotropic and lethal effects of ethanol and also increases its ability to cause physical dependency in rats (Nuzhnyi, Demechina, Zabirova, & Tezikov, 1999b).

TABLE 11.8. Chemical Composition of Homemade Beverages Samples: (A) Average for 80 Samples; (B) Typical Sample From the First Group; (C) Typical Sample From the Second Group; (D) Sample From the Third Group

	Concentration, mg/L			
Substance	A	B	C	D
Acetaldehyde	61.2 (5.3–849.8)	53.2	86.1	849.8
Substance RT 2.25	1.0 (0–12.9)	0	2.7	0
Acetone	3.8 (0–17.6)	4.4	2.2	13.7
Ethyl formate	1.0 (0–18.5)	0	1.6	17.5
Ethyl acetate	165.5 (3.3–655.3)	150.2	469.2	655.3
Methanol/isopropyl acetate	19.0 (0–655.3)	4.9	13.2	28.6
Isovaleric aldehyde	2.1 (0–36.5)	0	0.9	36.5
Diacetyl	2.4 (0–86.9)	0	2.6	86.9
Substance RT 5.05	0.4 (0–5.9)	0	1.2	5.9
sec-Butanol	11.9 (0–212.8)	0	0	0
Propanol-1	170.1 (8.2–566.4)	58.1	96.8	566.4
Isobutanol	1128.7 (18.7–3873.6)	870.2	1211.9	568.8
Isoamyl acetate	8.0 (0–236.9)	0	0	0
Butanol-1	14.5 (0–256.3)	0	0	33.5
Limonene	1.0 (0–21.5)	0	0	0
Isoamyl alcohol	1443.2 (35.5–4681.6)	981.2	927.1	1073.7
Substance RT 9.92 (acetoin)	18.3 (0–318.1)	11.6	6.8	318.1
Hexanol, ethyl lactate	9.8 (0–264.3)	3.1	4.9	12.5
Substance RT 11.82	1.2 (0–12.1)	1.6	1.0	0
Acetic acid	266.8 (0–763.6)	227.2	697.3	474.7
Furfurol	1.0 (0–11.1)	0	1.3	5.4
Propionic acid	2.6 (0–32.2)	0	7.3	13.3
Benzaldehyde	21.6 (0–999.1)	0	0	0
Isobutyric acid	89.6 (0–418.4)	51.0	105.8	10.5
Butyric acid	3.6 (0–36.5)	10.3	8.5	0
Ethylene glycol	1.8 (0–64.3)	0	0	64.3
Ethyl decanoate	5.4 (0–71.3)	0	0	0
Diethyl succinate	0.1 (0–9.3)	0	0	0
Isovaleric acid	28.2 (0–102.7)	17.7	29.7	0
Valeric acid	1.1 (0–11.8)	3.6	2.3	0
1,3-Propylene glycol	0.1 (0–5.5)	0	0	0
Phenylethyl alcohol	46.7 (0–153.8)	34.0	13.2	28.5
Dodecyl alcohol	5.6 (0–311.1)	0	0	0

Acetaldehyde, ethyl acetate, and other ether and aldehyde fraction components that sometimes are present in significant quantities are of little interest toxicologically. Ethers have little toxicity on their own, and acetaldehyde, due to its high reactive ability, reacts with food components and protein structures of gastric mucus, and it does not appear in pure form in the blood.

Higher spirits that are components of fusel oil (propanol, isobutanol, butanol, isoamyl alcohol, hexanol, etc.) are, like ethanol, classified as medium-toxicity components and do not differ much from ethanol in indicators of acute lethal effect. Concentrations in which higher spirits are found in industrial and noncommercial alcohol beverages cannot produce a pronounced impact on parameters of acute and subacute toxic effects of ethanol (Nuzhnyi, 1995; Nuzhnyi, Demechina, Zabirova, & Tezikov, 1999a). However, they do decrease the organoleptic qualities of beverages.

A Russian Federation study compared the acute and subacute toxicity of two *samogon* samples made from sugar and honey and solution of rectified ethanol of the same strength (Nuzhnyi, Savchuk, Demeschina, & Tezikov, 1999). An LD_{50} experiment on mice and rats found no differences between the ability of *samogon* and ethanol to cause physical alcohol dependency and subacute toxic effects (alcohol damage of heart or liver and gastric mucus). The only difference was related to *samogon* made from honey, which had a lesser psychotropic effect.

TOXICITY EXPRESS ANALYSIS CARRIED OUT IN THE RUSSIAN FEDERATION

We tested the toxicity of samples with an express method (testing subject cattle's sperm) that is used to evaluate vodka and ethanol toxicity (Eskov, Kajumov, & Nuzhnyi, 2000). Spermatozoon activity was registered. This activity usually decreases or increases in the presence of toxic substances. A sample was considered poisonous if its toxicity deviated unacceptably from the standard toxicity of ethanol solution. In total, we tested 66 samples.

According to toxicological research, 41 samples corresponded to vodka or 40% foodstuff ethanol solution ("Extra"), but 22 samples had a higher toxicity. These samples decreased spermatozoon motive activity (i.e., how quickly they were moving).

Table 11.9 shows the chemical composition of nontoxic and toxic homemade beverages. According to this table, there are no significant differences between toxic and nontoxic samples obtained by GC-FID and GC-MSD methods. Correlation analysis also did not reveal significant or logically valid connections between the concentration of substances and the toxicity of samples. These results showed that a homemade beverage's poisonous properties could be explained by the presence of other, nonvolatile, substances. Such substances might appear in alcohol beverages during different stages of their preparation or storage, or they could materialize during the fermentation process. It cannot be discounted that cattle spermatozoids have high sensitivity to the indicated hypothetical chemicals. To make the final judgment about the samples' toxicity, it is necessary to verify express-analysis results with the results of a full-scale toxicological study.

TABLE 11.9. Comparative Chemical Composition of (A) Nontoxic and (B) Toxic Homemade Beverages

	Concentration, mg/L	
Substance	A ($n = 41$)	B ($n = 22$)
Acetaldehyde	48.7 (9.2–91.9)	62.7 (5.3–126.6)
Substance RT 2.25	1.4 (0–12.9)	0
Acetone	2.5 (0–7.8)	6.6 (0–17.6)
Ethyl formate	1.3 (0–18.5)	0.5 (0–2.9)
Ethyl acetate	160.3 (5.2–631.6)	162.3 (3.3–290.4)
Methanol/isopropyl acetate	24.3 (0–655.3)	8.4 (0.2–23.6)
Isovaleric aldehyde	0.1 (0–3.4)	0
Diacetyl	0.8 (0-14.8)	0.7 (0–14.8)
Substance RT 5.05	0.1 (0–2.6)	0
sec-Butanol	10.4 (0–212.8)	16.3 (0–192.5)
Propanol-1	171.2 (56.7–470.7)	145.2 (8.2–349.0)
Isobutanol	1076.3 (269.9–3200.5)	1233.8 (18.7–3873.6)
Isoamyl acetate	9.5 (0–236.9)	6.2 (0–13.1)
Butanol-1	17.9 (0–256.3)	8.63 (0–30.5)
Limonene	0.1 (0–3.9)	2.9 (0–21.5)
Isoamyl alcohol	1429.8 (463.7–4661.6)	1478.8 (36.5–3169.0)
Substance RT 9.92 (acetoin)	2.8 (0–11.6)	2.2 (0–11.6)
Hexanol, ethyl lactate	12.4 (0–264.3)	5.3 (0–21.1)
Substance RT 11.82	1.2 (0–5.3)	1.5 (0–12.1)
Acetic acid	210.2 (0–697.3)	266.2 (0–763.6)
Furfurol	0.8 (0–11.1)	0.9 (0–7.2)
Propionic acid	3.1 (0–32.3)	1.8 (0–14.5)
Benzaldehyde	34.8 (0–999.1)	0
Isobutyric acid	83.7 (0–418.4)	98.4 (0–207.3)
Butyric acid	4.4 (0–26.4)	2.2 (0–36.5)
Ethyl decanoate	6.3 (0–71.3)	4.5 (0–33.4)
Diethyl succinate	0	0.4 (0–9.3)
Isovaleric acid	24.6 (9.0–98.5)	35.4 (0–102.7)
Valeric acid	1.53 (0–11.8)	0.5 (0–4.4)
1,2-Propanediol	0.1 (0–5.5)	0
Phenylethyl alcohol	43.8 (0–153.8)	50.9 (21.7–139.8)
Dodecyl alcohol	1.5 (0–61.0	14.4 (0–311.1)

EVALUATION OF ORGANOLEPTIC PROPERTIES

Tasting of *samogon* samples was carried out by the Central Tasting Committee of the Spirit and Liquor Industry, Russian Federation Ministry of Agriculture. There were 19 professional tasters at the committee session, and they selected 10 chemically significant differences from each of the other samples of *samogon* and 2 high-quality industrial beverages (Hennessy V.S. Cognac and Cutty Sark whisky). All the samples forwarded for tasting were depersonalized and randomly numbered. The evaluation process was based on a 10-point scale and took into account the appearance, scent, and taste of beverages. Evaluations of tasters were added and averaged. The highest estimate was 10 points.

Table 11.10 presents the tasting results and chemical composition. The highest evaluation was given to Henessy V.S. Cognac (8.37 points). The second place went to *samogon* number 10 (7.87 points), and the third to whisky (7.53 points). *Samogon* samples 2, 6, 11, and 12 did not differ markedly from whisky in taste and scent (7.32–7.49 points). The worst organoleptic features were found in samples 1, 5, 7, and 9 (6.43–6.76 points).

There was a significant interest to find out which volatile components influenced the taste and scent of the alcohol beverages. Correlation analysis revealed that high concentrations of isobutyric acid and diacetyl impacted negatively on organoleptic characteristics (Table 11.11). On the contrary, ethyl formate, ethyl decanoate, and furfurol concentrations positively correlated with tasters' high evaluation. The ability of isobutyric acid, diacetyl, ethyl formate, and ethyl decanoate to modify the taste of alcohol beverages is known to alcohol industry technologists. We saw no indication that furfurol could provide such an impact. Furfurol is a toxic ingredient and, according to the State Standard of the Russian Federation, its concentration in alcohol beverages must be regulated.

DISCUSSION

Results of the study showed that low-alcohol beverages and liquors made in different countries from various raw materials are similar to industrial beverages produced by means of fermentation and further distillation.

Detailed chemical analysis of 80 samples of *samogon* from the Russian Federation showed that half of them can be characterized as alcohol beverages of rather high quality (according to chemical criteria of quality that are applied to distilled alcohol beverages). The tasters' evaluation confirms this conclusion. Some 32.5% of the samples can be considered low-quality beverages with low organoleptic characteristics. Also, only 17.5% of the *samogon* samples had atypical chemical composition or contained very high amounts of volatile admixtures (very low quality). This prominent difference reflects differences in production conditions and the fact that standard, uniform technology is not used to produce homemade beverages.

About 30% of the *samogon* samples had high toxicity. However, the express method of toxicity estimation used in the study does not allow one to make definite statements about the toxicity of samples.

TABLE 11.10. Results of Tasting and Chemical Analysis of Samogon Samples, Cognac, and Whisky (Content of All Components, Except for Ethanol, are Presented in mg per Liter)

No. sample	1	2	3	4	5	6	7	8	9	10	11	12
				Cognac				Whisky				
Testing point	6.43	7.49	7.07	8.37	6.71	7.38	6.58	7.53	6.76	7.87	7.47	7.32
Ethanol % vol.	24.3	43.5	43.5	40.0	39.2	62.2	39.2	43.2	49.1	40.0	45.0	43.2
Acetaldehyde	76.4	60.2	37.9	36.9	59.13	31.05	25.4	30.4	63.18	61.83	63.72	22.08
Substance RT 2.55	0.7	0	4.2	0	0	0	5.04	0	0	4.48	7.58	2.32
Acetone	0	4.7	0	0	0	7.54	2.44	0	4.29	0	0	0
Ethyl formate	0	9.6	0	4.1	0	0	0	0	0	1.36	0	0
Ethyl acetate	44.6	631.6	37.8	184.5	94.18	331.2	36.6	134.0	208.8	197.2	110.08	86.2
Methanol/isopropyl acetate	2.76	19.1	655.3	166.2	3.1	12.19	7.91	61.0	4.025	5.4	1.72	3.68
Diacetyl	7.02	0	0	0	0	0	5.76	0	0	0	0	0
Propanol-1	65.9	86.1	56.7	137.6	68.8	359.6	87.6	281.7	151.9	94.9	124.18	93.2
Isobutanol	1036.7	290.7	686.14	616.0	799.5	3200.5	358.8	348.4	3873.6	642.04	1773.3	1064.8
Isoamyl acetate	0	0	0	3.7	0	26.7	0	8.8	10.49	0	0	0
Butanol-1	2.1	256.3	1.1	2.4	1.79	24.5	3.92	3.9	5.16	3.29	4.62	5.46
Isoamyl alcohol	963.6	542.9	524.8	1.5	513.8	4681.6	605.4	373.9	3169.1	679.6	1598.1	1394.5
Subst RT 9.92 (acetoin)	6.3	0	0	0	3.13	0	9.4	0	1.45	1.82	2.66	0
Hexanol ethyl lactate	1.9	264.3	11.7	67.8	1.58	3.84	15.36	3.7	2.4	4.01	1.76	3.88
Substance RT 11.82	0	0	1.05	0	1.28	1.9	0	0	2.43	2.35	1.55	1.3
Acetic acid	313.2	229.2	146.3	150.7	645.3	89.2	438.3	78.5	152.03	478.8	209.3	101.6
Furfurol	0	11.07	0.63	12.4	0	0	4.51	3.0	0	1.93	0	0
Propionic acid	3.12	0	32.2	0	0	0	17.94	0	0	5.53	3.38	0
Isobutyric acid	160	14.2	98.8	0	74.2	50.17	108.7	0	89.03	39.15	204.4	44.89
Butyric acid	2.4	0	17.3	0	0	0	7.0	0	0	14.05	9.96	5.88
Ethylene glycol	0	0	0	0	0	15.6	0	0	0	0	0	0
Ethyl decanoate	0	71.3	0	33.6	0	0	0	22.8	0	0	0	0
Isovaleric acid	30.4	24.1	27.4	10.9	10.24	24.0	34.8	0	0	16.06	40.96	15.5
Valeric acid	2.5	0	11.8	21.0	0	0	4.22	0	0	4.01	2.66	2.66
1,3-Propylene glycol	0	0	5.55	0	0	0	0	0	0	0	0	0
Phenylethyl alcohol	34.8	28.5	39.8	24.9	0	28.3	153.8	12.7	25.42	23.2	53.5	46.6
Enantic acid	0	0	0	0	16.2	0	0	0	0	0	0	0

TABLE 11.11. Results of Correlation Between Tasters' Evaluation of Alcohol Beverages (Samples of Samogon, Cognac, Whisky) and Content of Volatile Admixtures

Ingredient	*R** (*p*)	*T** (*p*)
Isobutyric acid	– 0.71 (.01)	– 0.53 (.02)
Diacetyl	—	– 0.46 (.04)
Ethyl formate	+ 0.67 (.02)	+ 0.54 (.02)
Ethyl decanoate	+ 0.62 (.03)	+ 0.49 (.03)
Furfurol	+ 0.62 (.03)	+ 0.47 (.02)

Note. *R**, Spearman rank coefficient correlation; *T**, Kendall correlation coefficient.

CONCLUSION

Throughout human history, people living in different geographical and climate zones have produced and used homemade alcohol beverages originating from various plants. Indeed, in most countries, homemade beverages are prototypes of industrial alcohol beverages. An exception relates to alcohol beverages made from rectified ethanol. Such beverages, such as vodka, contain very few admixtures and usually do not contain extractive substances.

The question of the toxicity of homemade alcohol beverages is very relevant in all the countries where such beverages retain their traditional popularity. Unfortunately, there are no data available about the relationship between the scale of noncommercial alcohol consumption and the medical, social, and alcohol-related demographic indicators in different countries. In particular, in the Russian Federation, excessive alcohol consumption is the main cause of both *samogon* and industrial alcohol poisonings. There is some documented evidence of acute poisonings from small amounts of *samogon* with toxic components, but this is the exception rather than the rule. Such components were deliberately added to the beverages in order to hide their defects (low alcohol content) and to increase their intoxication effect. (Usually, *samogon* intended for sale was infused with tobacco leaves, hop cones, or chicken droppings.) There exists no reliable data about the high chronic toxicity of *samogon* and its supposed promotion of alcoholism and alcohol-related somatic diseases.

It is often assumed that moonshine beverages are inherently dangerous. However, the results of the analyses described in this chapter indicate that the samples do not pose a toxic hazard and, in many cases, the alcohol beverage is of high quality in terms of taste. Although the number of samples actually analyzed in some of the countries was very small, there is no reason to suppose that they were atypical.

There is the possibility that laboratories in some developing countries may not be able to achieve the same standard of quality control in their estimations as has been reported from Russia, but this is a matter of speculation and not of fact. Certainly more research is necessary and the findings should still be looked upon as tentative. However, it seems likely that many myths about the alleged dangers of consuming these beverages may well be overturned.

REFERENCES

Bechterev, V. M. (1913). About alcohol sanitation. (Бехтерев В.М. *Об алкогольном оздоровлении.* В кн.: Вопросы алкоголизма. Вып. 1. СПб, 1913, 81–100.)

Boras, E., Coutelle, C., Rosell, A., & Parres, J. (2000). Genetic polymorphism of alcohol dehydrogenase in Europeans: The ADH2*2 allele decreases the risk for alcoholism and is associated with ADH3*1. *Hepatology, 31,* 984–989.

Brechman, I. I., & Kublanov, M. G. (1983). Concept of structural information in pharmacology and nutrition science. (Брехман И.И., Кубланов М.Г. *Концепция структурной информации в фармакологии и науке о питании.* Владивосток, ДВНЦ АН СССР, 1983, 28 с.).

Edenberg, H. J.,& Bosron, W. F. (1997). Comprehensive toxicology. In I. G Sipes, C. A. McQueen, & A. J Gandolfi (Eds.), *Biotransformation,* Vol. 3, *Alcohol dehydrogenases* (pp. 119–131). New York: Pergamon.

Eskov, A. P., Kajumov, P. I., & Nuzhnyi, V. P. (2000). Evaluating toxicity of spirits, spirits' solutions and vodkas by in vitro method using sperm of cattle as a test object. (Еськов А.П., Каюмов Р.И., Нужный В.П. Оценка токсичности спиртов, спиртовых растворов и водок методом in vitro с использованием спермы крупного рогатого скота в качестве клеточного тест-объекта. *Токсикол. вестник,* 2000, *5,* 16–21.)

Goedde, D., & Agrawal, D. P. (1992). Distribution of ADH2 and ALDH2 genotypes in different populations. *Hum. Genet., 88,* 344–346.

Horwitz, W. (Ed.). (1975). *Official methods of analysis of the Association of Official Analytical Chemists* (12th ed.). Washington: AOAC.

Ishibashi, T., Harada, S., Fujii, C., & Ishii, T. (1999). Relationship between ALDH2 genotypes and choice of alcohol beverages. *Nihon Arukoru Yakubutsu Igakkai Zasshi, 34*(2), 117–129.

Ivanets, N. N., & Koshkina, E. A. (2000). Medical and social consequences of alcohol abuse in Russia. (Иванец Н.Н., Кошкина Е.А. Медико-социальные последствия злоупотребления алкоголем в России. *Новости науки и техники. Сер. Мед. Вып. Алкогольная болезнь.* ВИНИТИ, 2000, *1,* 3–8.)

Kilonzo, G. P., & Pitkanen, Y. T. (1992). *POMBE report of alcohol research project in Tanzania 1988–90.* Helsinki: University of Helsinki, Institute of Development Studies.

Kulapina, T. I. (1998). Alcohol use in Russia and developed countries within the last 150 years (1856–1996). (Кулапина Т.И. Потребление алкоголя в России и развитых странах за последние полтора века (1856-1996). *Известия Российской Академии Наук. – Сер. Географич.,* 1998, *4,* 6–20.)

Kunze, W. (1998). *Technologie Brauer und Malzer.* Berlin: VLB.

Lai, J., Kumar, C. V., Suresh, M. V., & Indira, M. (2001). Effect of exposure to a country liquor (*toddy*) during gestation on lipid metabolism in rats. *Plant Foods Hum. Nutr., 56*(2), 133–143.

Meier-Tackmann, D., Leonhardt, R. A., Agarwal, D. P., & Goedde, H. W. (1990). Effects of acute ethanol drinking on alcohol metabolism in subjects with different ADH and ALDH genotypes. *Alcohol, 7,* 413–418.

Nuzhnyi, V. P. (1995). Toxicological characteristics of ethanol, alcohol beverages and admixtures that they contain. (Нужный В.П. Токсикологическая характеристика этилового спирта, алкогольных напитков и содержащихся в них примесей. *Вопр. наркологии,* 1995, *3*, 65–74.)

Nuzhnyi, V. P. (1997). Beer: Chemical composition, nutritional value, biological activity and consumption. (Нужный В.П. Пиво: химический состав, пищевая ценность, биологическое действие и потребление. *Вопр. наркологии,* 1997, *4,* 68–76.)

Nuzhnyi, V. P., & Prichozhan, L. M. (1996). New approach to the problem of alcohol beverages' toxicity. (Нужный В.П., Прихожан Л.М. Новый взгляд на проблему токсичности алкогольных напитков. *Токсикол. вестник*, 1996, *5,* 9–16.)

Nuzhnyi, V. P., Demechina, I. V., Zabirova, I. G., & Tezikov, E. B. (1999a). Influence of fusel oil, ether and aldehyde fraction on acute toxicity and psychotropic activity of ethanol. (Нужный В.П. и соавт. Влияние компонентов сивушного масла и эфироальдегидной фракции на острую токсичность и наркотическое действие этилового спирта. *Токсикол. вестник*, 1999, *2,* 2–8.)

Nuzhnyi, V. P., Demechina, I. V., Zabirova, I. G., & Tezikov, E. B. (1999b). Studying the impact of bethanol on acute and subacute effects of ethanol. (Нужный В.П. и соавт. Исследование влияния мктанола на острое и подострое токсическое действие этилового спирта. *Вопр. наркологии,* 1999, *3*, 54–59.)

Nuzhnyi, V. P., Savchuk, C. A., Demeschina I. V., & Tezikov E. B. (1999). Composition and toxicity of *samogon* made of sugar and honey. (Нужный В.П. и соавт. Состав и токсичность самогонов из сахара и меда. *Реф. сб. ВИНИТИ: Новости науки и техники. Сер. Медицина. Вып. Алкогольная болезнь,* 1999, *6,* 1–10.)

Ogurtsov, P. P. (1989). Hidden population health and Russian Federation healthcare budget losses due to chronic alcohol intoxication (alcohol disease). (Огурцов П.П. Скрытые потери здоровья населения и бюджета здравоохранения РФ от хронической алкогольной интоксикации (алкогольной болезни). *Новости науки и техники. Сер. Мед. Вып. Алкогольная болезнь.* ВИНИТИ, 1998, *6,* 8–20.)

Ogurtsov, P. P., Garmash, I. V., Miandina, G. I., & Moiseev, V. S. (2001). Alcohol dehydrogenase ADH2-1 and ADH-2 allelic isoforms in the Russian population correlate with type of alcohol disease. *Addiction Biology, 6,* 377–383.

Pasini, G. (1992). The composition of alcoholic beverages. *Alcologia, 4*(2), 137–142.

Sarin, S. K., Malhotra, V., Nayyar, A., & Broor S. L. (1988). Profile of alcohol liver disease in an Indian hospital. A prospective analysis. *Liver, 8*(3), 132–137.

Savchuk, S. A., Brodski, E. S., Formanovski, A. A., & Rudenko, B. A. (1999). Gas chromatography and chromo-mass spectric detection of glycols in drinking water and liquors. (Савчук С.А. Бродский Е.С., Формановский А.А., Руденко Б.А. Газохроматографическое и хромато-масспектрометрическое определение гликолей в питьевой воде и спиртных напитках. *Журн. Аналит. Хим.*, 1999, *8,* 131–139.)

Savchuk, S. A., Vlasov, V. N., Appolonova, S. A., & Grigorian, B. G. (2001) Using chromatography and spectrometry for identification of counterfeit alcohol beverages. (Савчук С.А. и соавт. Применение хроматографии и спектрометрии для идентификации подлинности спиртных напитков. *Журн. Аналит. Хим.*, 2001, *3,* 246–264.)

Shatirishvili, I. M. (1998). Chromatography in enology. (Шатиришвили И.М. *Хроматография в энологии.* Тбилиси: Ганатлеба, 1998, 110 с.)

Sibachov, T. S. (2001). Basics of Cognac production technology in Russia. (Сибахов Т.С. *Основы технологии коньячного производства России.* Новочеркасск, 2001, 160 с.)

Suchodolova, G. N., Strachov, S. I., & Chundoeva, S. S. (2000). Distinguishing features of acute poisonings among children. (Суходолова Г.Н., Страхов С.И., Хундоева С.С. Злоупотребление алкоголем в России и здоровье населения. Информационные материалы и рекомендации для врачей. *Особенности острых отравлений у детей* (с. 107–112). М.: Лабпресс. Российская ассоциация общественного здоровья, 2000.)

Waterhouse, A. L., & Frankel, E. N. (1993). *Wine antioxidants may reduce heart disease and cancer. Proc.* OIV 73rd General Assembly, San Francisco, August 29-September 3, 1993. Paris: OIV (1–15).

Yamauchi, M., Maezawa, Y., Toda, G., Suzuki, H., & Sakurai, S. (1995). Association of a restriction fragment length polymorphism in the alcohol dehydrogenase 2 gene with Japanese alcohol liver cirrhosis. *Journal of Hepatology*, *23,* 519–523 (b).

Yin, S. J., Liao, C. S., Chen, C. M., & Lee, S. C. (1992). Genetic polymorphism and activities of human lung alcohol and aldehyde dehydrogenases: Implications for ethanol metabolism and cytotoxity. *Biochem Genetics, 30,* 203–215.

Yokoyama, A., Muramatsu, T., Omori, T., & Ishii, H. (1999). Alcohol and aldehyde dehydrogenase gene polymorphisms influence susceptibility to esophageal cancer in Japanese alcohols. *Alcohol: Clinical & Experimental Research, 23*(11), 1705–1710.

Zariwala, M. B., Lalitha, V. S., & Bhide, S. V. (1991). Carcinogenic potential of Indian alcohol beverages (country liquor). *Indian Journal of Experimental Biology, 29*(8), 738–743.

Zariwala, M. B., Kayal, J. J., & Bhide, S. V. (1993). Effect of country liquor (Indian alcohol beverage) on carcinogen activating and detoxifying enzymes. *Indian Journal of Physiology Pharmacology, 37*(1), 85–87.

Index